AF801681

Geological Genetic Analysis and Application of Oil and Gas Reservoirs

Huanqing Chen

Geological Genetic Analysis and Application of Oil and Gas Reservoirs

石油工业出版社
PETROLEUM INDUSTRY PRESS

Springer

Huanqing Chen
Department of Oil Field Development
Research Institute of Petroleum
Exploration & Development
PetroChina Company Limited
Beijing, China

ISBN 978-981-97-7627-6 ISBN 978-981-97-7628-3 (eBook)
https://doi.org/10.1007/978-981-97-7628-3

Jointly published with Petroleum Industry Press
The print edition is not for sale in China (Mainland). Customers from China (Mainland) please order the print book from: Petroleum Industry Press.
ISBN of the Co-Publisher's edition: 978-7-5183-4817-6

© Petroleum Industry Press 2024

This work is subject to copyright. All rights are solely and exclusively licensed by the Publisher, whether the whole or part of the material is concerned, specifically the rights of reprinting, reuse of illustrations, recitation, broadcasting, reproduction on microfilms or in any other physical way, and transmission or information storage and retrieval, electronic adaptation, computer software, or by similar or dissimilar methodology now known or hereafter developed.
The use of general descriptive names, registered names, trademarks, service marks, etc. in this publication does not imply, even in the absence of a specific statement, that such names are exempt from the relevant protective laws and regulations and therefore free for general use.
The publishers, the authors, and the editors are safe to assume that the advice and information in this book are believed to be true and accurate at the date of publication. Neither the publishers nor the authors or the editors give a warranty, express or implied, with respect to the material contained herein or for any errors or omissions that may have been made. The publishers remain neutral with regard to jurisdictional claims in published maps and institutional affiliations.

This Springer imprint is published by the registered company Springer Nature Singapore Pte Ltd.
The registered company address is: 152 Beach Road, #21-01/04 Gateway East, Singapore 189721, Singapore

If disposing of this product, please recycle the paper.

Preface

Geological genetic analysis of oil and gas reservoirs in field development analyzes the genesis of reservoirs from a geological perspective. It identifies the primary factors controlling the genesis using various data (e.g. geology, geophysics, geochemistry, dynamic monitoring, and production performance), thereby providing reference and basis for reservoir characterization and implementing a development (adjustment) plan. During field development, especially in the middle to late stage, most researches focus on qualifications using computer technology and mathematical/physical methods but neglect the geological genetic analysis of structures, reservoirs, and OGW systems. Such neglection has had severe consequences in practical activities because many problems in field development are related to geological genesis involving several factors. Deficiencies in these studies have restricted field development and oil/gas recovery. Also, a viewpoint exists that geological research is insignificant because the underground resources will be recovered if development activities are performed. This viewpoint is, indeed, wrong. Qiu Yinan, a Chinese oilfield development expert, indicated that an oilfield must be developed after it is well understood, and practices in many oilfields abroad have proved the passiveness of oilfield development due to unknown geological characteristics and unclear information about oil zones, typically analyzed on a macro, rough, or general scale. Genetic analysis is critical to correctly understanding the geological characteristics of oil and gas reservoirs. This book presents a systematic description of the geological genetic analysis of oil and gas reservoirs regarding inputs, methods, problems, and applications to provide reference for relevant research.

This book includes five chapters: Introduction, Geological Genetic Analysis of Structures, Geological Genetic Analysis of Reservoirs, Geological Genetic Analysis of OGW Systems, and Application of Geological Genetic Analysis of Oil and Gas Reservoirs in Field Development. According to the data sources for the geological genetic analysis of oil and gas reservoirs, the research is conducted in three aspects: structures, reservoirs, and OGW systems. From the research status at home and abroad, the research items and methods of the geological genetic analysis of oil and gas reservoirs are summarized, the geological genetic analysis of special oil and gas reservoirs is outlined, and the problems and development directions of geological genetic analysis of oil and gas reservoirs are proposed. The research methods for the geological genetic analysis of oil and gas reservoirs include field outcrop and modern sediment investigations, core observation and description, thin section observation and identification, geophysical prediction and interpretation, geostatistics, analytical tests, and physical and numerical simulations, and reservoir dynamic monitoring and production performance analysis, described with their advantages and disadvantages. The geological genetic analysis of oil and gas reservoirs has six problems. (1) It is insufficiently concerned with field development. (2) The research methods tend to be qualitative. (3) Geological genetic analysis and reservoir characterization are inadequately combined. (4) The precision of reservoir characterization restricts the accuracy of geological genetic analysis. (5) The geological genetic analysis of oil and gas reservoirs is incomprehensive. (6) Many challenges exist in the geological genetic analysis of special oil and gas reservoirs. The research on the geological genetic analysis of oil and gas reservoirs will focus on its role in addressing the field development problems, quantification-oriented improvement by various simulation methods, enhanced contribution to reservoir characterization, advancement using reservoir characterization, extended application in field development, and application for special oil and gas reservoirs. This book elaborates the geological genetic analysis of oil and gas reservoirs regarding structures, reservoirs, and OGW systems, using the volcanic gas reservoir in the first member of Yingcheng Formation in the Xudong area, the Songliao Basin, the alluvial-fan glutenite oil reservoir in the Lower Karamay Formation in an area at the Junggar Basin's northwestern margin, and a steam-flooded heavy oil reservoir in the Yulou reservoir in the Liaohe Basin's West Sag as examples. Moreover, this book systematically introduces the application of geological genetic analysis of oil and gas reservoirs in field development from some aspects, including fine stratigraphic division based on high-resolution sequence stratigraphy, control/influence of sedimentary genetic types on the architecture of glutenite reservoirs, lithofacies characteristics of volcanic gas reservoirs and their influence on reservoir physical properties, the geological genesis of flow barrier in heavy oil reservoirs with thermal recovery, geological modeling of sedimentary microfacies based on geological genetic analysis in fine reservoir description, and geobody genesis classification and evaluation in heavy oil reservoirs with thermal recovery.

This book incorporates the achievements of research projects where the author worked as a primary participant, including the Subject 1–Safe Development and CO_2 Utilization Technologies for CO_2-containing Gas Reservoirs under Project 16 "Development of Large-scale Oil and Gas Fields and Coalbed Methane" (National Science and Technology Major Project in the 11th Five-Year Plan), the fine reservoir description project of PetroChina Company Limited, and practical production projects in Daqing volcanic rock, Xinjiang glutenite and Liaohe heavy oil reservoirs.

We are particularly grateful for the guidance and assistance provided by Prof. Hu Yongle, Prof. Shi Chengfang, Prof. Jin Jiuqiang, Prof. Tian Changbing, Prof. Li Baozhu, Prof. Gao Xingjun, Prof. Zhao Yingcheng, Prof. Ran Qiquan, Prof. Zhang Hujun and other experts from PetroChina Research Institute of Petroleum Exploration & Development! We also appreciate the guidance and help of Dr. Yan Lin, Dr. Tong Min, Dr. Wang Yongjun, and Dr. Zhang Jing of the Daqing Volcanic

Rocks Research Team, Dr. Li Shunming and Dr. Jiang Ping of the Xinjiang Glutenite Research Team, Engineer Wang Jue, Engineer Du Yijing, and Engineer Yao Yao of the Liaohe Heavy Oil Research Team, and Hong Yao, Sui Yuhao and Bie Hanyu of the Fine Reservoir Description Research Team.

We would like to express our heart-felt thanks to the directors and experts from Daqing, Xinjiang, Liaohe, Changqing, Jilin, Dagang, Jidong, Huabei, Qinghai, Yumen, Tuha, Tarim, Southern and other oilfields of PetroChina for their grand support and assistance to this work. Our special thanks also go to Hu Haiyan, Wang Liangang, Wu Hongbiao and Cao Chen (the Deputy Chief Geologist, the Director General, the Senior Company Expert, and the Senior Executive, respectively, of PetroChina Oil & Gas & New Energy Branch Company).

Due to the limited level of knowledge and experience, we could not avoid inappropriate statements in this book. Your comments and criticism are thereby warmly welcomed.

Beijing, China Huanqing Chen

Contents

Chapter 1
Introduction

An oil/gas field is the sum of all the oil and gas reservoirs in the same area controlled by a single local structural unit (Zhang et al. 2007). An oil and gas reservoir is an accumulation of oil and gas in a single trap. A reservoir is the basic unit of oil and gas accumulation in the crust, with a unified pressure system and oil–gas–water (OGW) systems. In the development of oil/gas fields, the geological genetic analysis of oil and gas reservoirs is mainly based on various data (e.g. geology, geophysics, geochemistry, dynamic monitoring, and production performance) to analyze the genesis of oil and gas reservoirs from a geological perspective, determine the main factors controlling the genesis of oil and gas reservoirs, and provide a reference and basis for the implementation of various measures in reservoir characterization and development (adjustment) plans (Chen et al. 2019a, b). Geological genetic analysis underpins the whole process of oilfield development and is needed for stratigraphic division and correlation, sedimentary facies research, reservoir evaluation, geological modeling, flow unit research, etc. The precision and accuracy of the analysis directly affects the precision and accuracy in understanding oil and gas reservoirs; therefore, geological genetic analysis has attracted the attention of more and more oil/gas field developers.

1.1 Research Status in China and Abroad

Research on the geological genesis of oil and gas reservoirs involves many aspects, such as structures, reservoirs, and OGW systems. Many researchers in China and abroad have conducted studies in this domain (Malone et al. 2002; Lobach-Zhuchenko et al. 2005; Hogan et al. 2007; Zhu 2008a; Sarmah et al. 2010; Zeng et al. 2010; Huang et al. 2012; Agha et al. 2013; Gao et al. 2013b; Yuce et al. 2014; Furukawa et al. 2014; Li et al. 2015a; Chen 2015; Xu et al. 2017; Zhang et al. 2018; Chen et al. 2019b). Malone et al. (2002) reported the variable methane fluxes in shallow marine systems over geologic time through analyzing the composition

© Petroleum Industry Press 2024

H. Chen, *Geological Genetic Analysis and Application of Oil and Gas Reservoirs*,
https://doi.org/10.1007/978-981-97-7628-3_1

and origin of pore waters and authigenic carbonates on the New Jerseyshelf, and they concluded that methane flux likely plays an important role in long-term carbon cycling on passive continental margins. Lobach-Zhuchenko et al. (2005) studied the geological setting, geochemical characteristics and implications for the origin of the Archean snukitoid series of the Baltic Shield. Hogan et al. (2007) analyzed the geologic origin of salinization in semi-arid river, that is, the role of sedimentary basin brines, and concluded that the geologic salinity source of groundwater might be more effective in reducing salinity than changes in agricultural practices. Zhu (2008a, b) analyzed the geologic genesis of high oil saturation in low-porosity and low-permeability reservoirs by taking the reservoir in the Dongying Depression in the Shengli Oilfield for example, and concluded that the main factors affecting oil saturation include reservoir dynamics, reservoir pore structures, and deposition of overlying strata after reservoir formation. Sarmah et al. (2010) discussed the different geological origins of asphaltenes separated from crude oils in light of interfacial tension and thermogravimetric analysis. Zeng et al. (2010) investigated the origin and geological significance of cross fractures in the Upper Triassic Yanchang Formation in the Ordos Basin, China, through field outcrop analysis, core observation, and laboratory experiment, and indicated that cross fractures are the main pathway for fluid movement in the process of water injection and have great impact on oilfield development. Huang et al. (2012) presented the geologic origin of aquifer with high resistivity in the carbonate reservoirs in South China through a case study from the Lower Permian Qixia Formation in western Sichuan Basin. Agha et al. (2013) analyzed the geologic origin of Egyptian bentonitic clays using the clay mineral deterministic prediction methods. Gao et al. (2013a, b) investigated the microscopic geologic origin of ultra-low permeability reservoir of the Yanchang Formation in the Ordos Basin, and concluded that high cement content, high chlorite content and complex form, halite content, high ferrocalcite content, illite cement upgrowth, and bad lomontite cement corrosion are the key factors leading to poor physical properties and complex pore structure of ultra-low permeability reservoir. Yuce et al. (2014) reported the origin and interactions of fluids circulating over the AmikBasin (Hatay, Turkey) and relationships with the hydrologic, geologic and tectonic settings. Furukawa et al. (2014) studied the origin and mode of lithic-rich breccias in Aso Volcano, Japan, from the aspects of geological, paleomagnetic and petrological reconstruction. Li et al. (2015a, b) identified the implicit well logging geological information by interpreting the genetic mechanism, accumulation mode and formation background of rock minerals through geological attribute analysis of well-logging information using genetic relationships, difference comparisons, and other methods. Chen (2015) investigated the geologic origin of calcareous mudstone in the Nantun Formation of the Beier Sag and its petroleum geological significance, and concluded that the calcareous mudstone the Nantun Formation has high GR, high RT, high AC, and low density logging responses, and exhibits continuous strong reflections on the seismic sections. He also indicated that the Nantun Formation was formed in a vast lake basin, brackish and deep water sedimentary environment, with stable and broad deposition, and it is an important marker bed for stratigraphic correlations in the Beier Sag. Xu et al. (2017) identified the gas hydrates in near seafloor sediments in the eastern part of the Pearl River Mouth Basin based on seismic data and analyzed their geological origin. Zhang et al. (2018) analyzed the fractal characteristics of the micropore structure of sandstone reservoirs based on geological origin with the Daqing placanticline in the Songliao Basinas an example. Chen et al. (2019b, a) carried out a multi-parameter quantitative evaluation of carbonate reservoirs based on geological genetic analysis and concluded that the comprehensive evaluation results of reservoirs in a single well are positively correlated to the actual productivity.

To sum up, the studies on geological origin of oil and gas reservoirs in China and abroad have used the data primarily covering geology, geophysics, geochemistry, test, reservoir dynamic monitoring, and production performance, etc. In terms of research items, the studies abroad focus on structures, especially faults, and the distribution of oil, gas, and water, whereas the studies in China focus on reservoirs and the distribution of oil, gas, and water. Specifically, in the studies on reservoirs, the sedimentary environments and genetic models of reservoir formations are mainly analyzed from the perspective of sedimentology; in the studies on the distribution of oil, gas, and water, the origins of low-resistivity oil zones or low-porosity and low-permeability reservoirs are mainly treated. The target rock types include clastic rocks, carbonate rocks and volcanic rocks. Currently, the geological genetic analysis of oil and gas reservoirs has played a great role in the development of oil/gas fields. However, with the advancements in computers and other related technologies and methods, the continuous pursuit of semiquantitative and quantitative research methods in oil/gas field development has gradually weakened the practice of geological genetic analysis. This trend also directly leads to a consequence that many problems encountered in the development and research of oil/gas fields cannot be correctly understood and explained, which seriously affects the improvement of oil/gas field development. Therefore, it is necessary to comb through and summarize the research related to the geological genetic analysis of oil and gas reservoirs, so as to attract the attention of more researchers and provide references for the efficient development of oil and gas reservoirs.

1.2 Data Sources for Geological Genetic Analysis of Oil and Gas Reservoirs

The geological genetic analysis of oil and gas reservoirs involves various inputs and research methods; hence, the types of data used in such research are also very rich, including field outcrops, modern sediments, cores, thin sections, well logs, seismic survey, tests, simulation experiments, reservoir dynamic monitoring, and oil/gas field development. When analyzing the geological origin of alluvial-fan glutenite reservoirs in an area at the Junggar Basin's northwestern margin, I mainly used geologic, core and logging data. To be specific, I determined the types of sedimentary facies and subfacies through core observation and description, statistical analyses, and comparisons of well-logging curve shapes, in combination with the regional geological setting, and classified the sedimentary microfacies and reservoir architecture based on the analysis of electrofacies (Fig. 1.1). I divided the fourth-order architectures of the target reservoirs into 13 types: channel-flow gravel, channel-shoal sand/gravel, sheetflood inner sand, sheetflood inner fine grains, laminar-flow gravel, sheetflood outer sand, sheetflood outer fine grains, braided channel, braided sand/gravel bar, cross-flow sand, cross-flow fine grain, runoff channel, and interchannel fine grain. This is an example of geological genetic analysis of reservoirs in the geological genetic analysis of oil and gas reservoirs. Because a reservoir is the result of the comprehensive action of many factors, it should be analyzed comprehensively by combining various data to avoid ambiguity. Similarly, multiple factors should also be integrated in the geological genetic analysis of structures and OGW systems. Although the data of field outcrops, modern sediments, and cores can provide the most intuitive information for geological genetic analysis, they are mainly used in qualitative research to help researchers establish various genetic models. To conduct quantitative geological genetic analysis, the effective information should be extracted from logging, seismic, test, reservoir dynamic monitoring, and production performance data.

1.3 Research Items of Geological Genetic Analysis of Oil and Gas Reservoirs

The geological genetic analysis of oil and gas reservoirs involves a wide range of items that mainly include the following three aspects.

(1) Geological genetic analysis of structures

For example, in structure interpretation, geological genetic analysis is made on faults, fractures, and microstructures. In the multi-information fracture characterization of volcanic reservoirs of the first member of Yingcheng Formation (hereinafter referred to as "Ying-1 Member") in the Xudong area of the Songliao Basin, Chen et al. (2016a, b) classified the fractures by genesis and discussed the scale of fractures, rock types, and the data basis for identification of fractures.

(2) Geological genetic analysis of reservoirs

For example, in the study of sedimentary facies, the microfacies are determined based on the rock types, sedimentary structures, mudstone color, biological fossils, grain-size analysis, and sedimentary cycles. In the study of pore structure, geological genetic analysis is made to determine whether the pore structure of a reservoir is dominated by primary pores or secondary pores, or a combination of both. In the study of pore space, geological genetic analysis is conducted to determine whether the pore space is composed of pores, fractures, vugs, or others. In the comprehensive quantitative evaluation of reservoirs, geological genetic analysis should be done, as the first step, to determine the geological origin of the reservoirs, and then appropriate

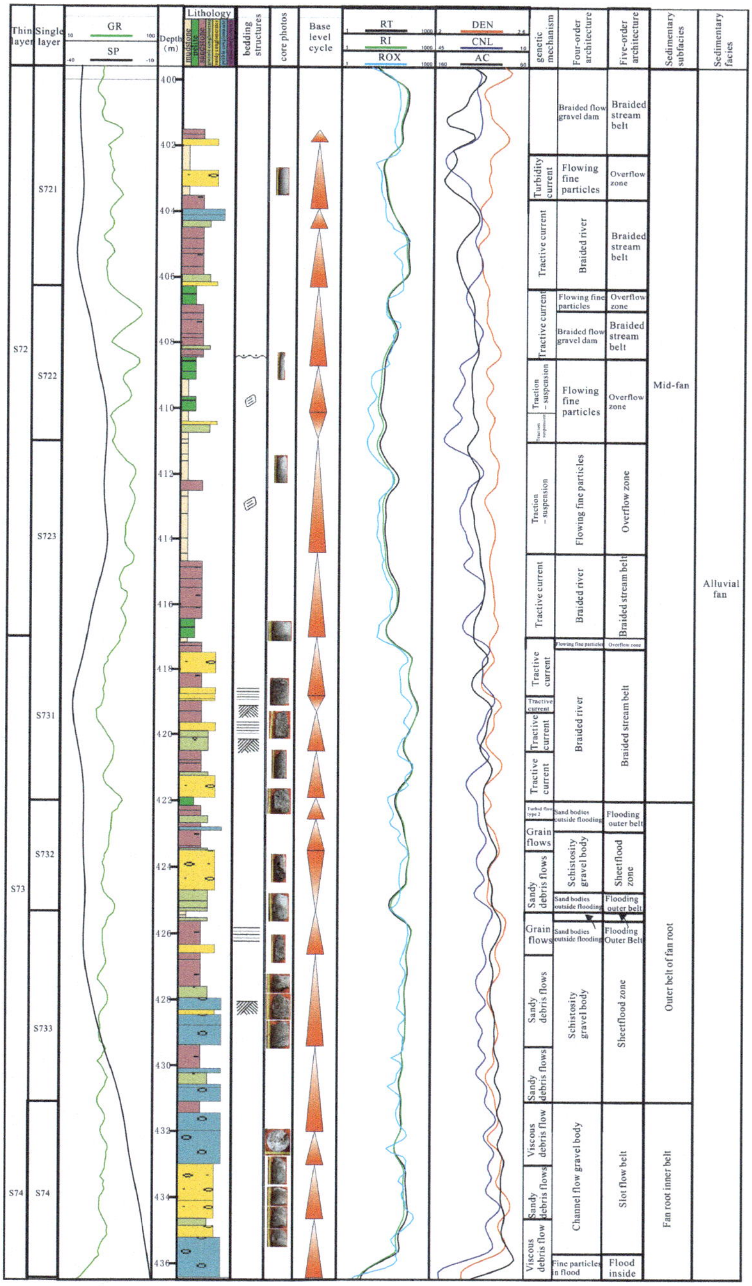

Fig. 1.1 Comprehensive lithology histogram of single well of J4 well in conglomerate reservoir of Kexia Formation in a certain area of northwestern margin of Junggar Basin

classification and evaluation parameters are selected for the quantitative evaluation of reservoirs. It can be imagined that if the procedure of geological genetic analysis of reservoirs is missing, the quantitative evaluation of reservoirs may be inaccurate due to errors in parameter selection and even be a meaningless number-oriented work. In a quantitative evaluation of beds in the Yulou reservoir in an area of the Liaohe Basin's West Sag, Chen et al. (2017) first analyzed the geological genesis of reservoir beds and then selected five parameters, i.e. shale content, porosity, permeability, reservoir thickness, and permeability variation coefficient, which fully reflect the influencing factors of the reservoirs, for purpose of comprehensive quantitative evaluation of reservoirs.

(3) Geological genetic analysis of OGW systems

If there are uniform gas–oil, oil–water, or gas–water contacts in a hydrocarbon accumulation, it means that the reservoir systems are interconnected in the process of reservoir forming, which are considered as an OGW system. The distribution and occurrence of OGW systems are directly related to the reserves estimation and development deployment; therefore, they are very important research topics in the development of oil/gas fields (Qiu et al. 1996). The geological genetic analysis of an OGW system is performed to understand the distribution characteristics of fluids (oil, gas, and water) in the reservoir. Kotarba et al. (2014) discussed the origin of gaseous hydrocarbons, noble gases, carbon dioxide and nitrogen in the Carboniferous and Permian strata of the distal part of the Polish Basin by using geological and isotopic approach. Figure 1.2 shows the formation pressures of a well in the Yulou reservoir in an area of the Liaohe Basin's West Sag, which was plotted from the repeat formation test (RFT) data. It can be seen that the formation pressure of the well changes greatly at the depth of 1060 m, suggesting that there is an oil–water contact (OWC) near this depth, above and below which the pressure systems are different. The same profiles are obtained for other two wells from RFT data. Thus, it is preliminarily determined that the OWC of the Yulou reservoir is located at approximately 1060 m. Combined with the geological setting and structural evolution history, it is confirmed that the Yulou reservoir belongs to a structural–lithologic oil reservoir.

1.4 Research Methods for Geological Genetic Analysis of Oil and Gas Reservoirs

Depending on the research purpose and data availability, a variety of methods can be used in the geological genetic analysis of oil and gas reservoirs, including field outcrop and modern sediment investigations, core observation and description, thin section observation and identification, geophysical prediction and interpretation, geostatistics, analytical tests, and physical and numerical simulations.

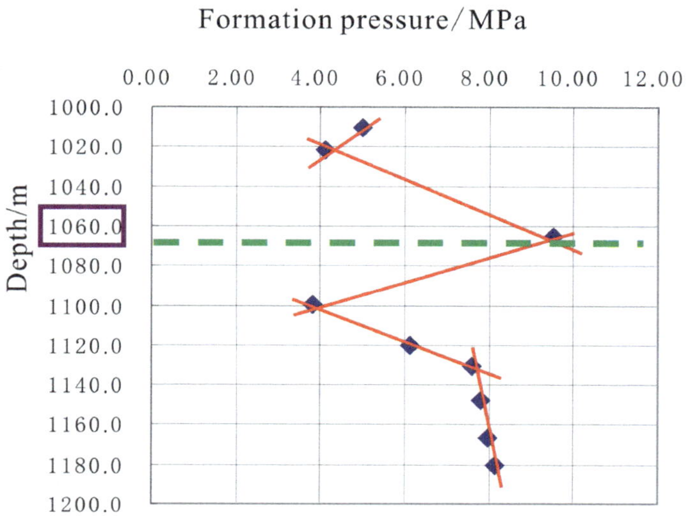

Fig. 1.2 Characteristics of stratum pressure profile of Yulou oil bearing sets in some steam flooding experimental area in the West depression of Liaohe Basin (RFT data, True vertical depth)

1.4.1 Field Outcrop and Modern Sediment Investigations

Field outcrop and modern sediment investigations represent one of the common research methods for geological genesis of reservoir. Field outcrop investigation involves the analysis of geological phenomena (e.g. faults, lithologies, sedimentary structures, and biological fossils) of field outcrops that have similar geological characteristics to the target of interest in order to understand the development characteristics and models of structures and reservoirs in the target, thereby enabling the recognition of geological genetic elements of reservoirs to guide the effective development of the reservoir. Modern sediment investigation focuses on sedimentary phenomena (e.g. rivers, lakes, and deltas) in areas with identical or similar sedimentary facies types to the target of interest, so as to get more understanding of the sedimentary processes and products to provide reference for analyzing the genesis of the oil reservoir. In China, there are abundant field outcrop and modern sediment data in many petroliferous basins, such as Songliao, Ordos, Sichuan, and Junggar, which allow systematic geological genetic analysis of oil and gas reservoirs (Fig. 1.3). Geological genetic analysis of oil and gas reservoirs using field outcrops and modern sediments mainly follows the principle of uniformitaranism (Feng 1994). It should be noted that comparison is necessary for the geological genetic analysis of oil and gas reservoirs using field outcrops and modern sediments, that is, the target oil and gas reservoirs are compared with field outcrops and modern sediments in light of basic geological characteristics, in order to confirm that the geological genetic models summarized from the field outcrops and modern sediments can be applied, after revision if necessary, to the target reservoirs. This method is intuitive and can easily build geological genetic model. If conditions permit, geological sections of field outcrops can be measured to obtain sufficient quantitative data for purpose of semiquantitative analysis of geological genetic model. However, field outcrops and modern sediments are geologically different from underground oil and gas reservoirs to a certain extent, so analogy analysis is essential, and the resulting geological genetic model may deviate from reality.

1.4.2 Core Observation and Description

Core observation and description is an important research method for geological genetic analysis of oil and gas reservoirs. As the most prominent advantage, it can visually describe the most basic features such as structures and reservoirs. This method can help understand the phase and scale of structures/reservoirs, fractures, lithology, sedimentary structures, grain size, roundness and sorting of rock particles, mudstone color, paleontological fossils, and spatial distribution of reservoirs. I used core data in studying the sedimentary facies of the Yulou reservoir in an area of the Liaohe Basin's West Sag. The core data intuitively displayed the presence of channel-lag sediments, trough cross bedding, horizontal bedding, carbonized plant debris on bedding surfaces, and gastropod and snail fossils, in the Yulou reservoir, by which fan delta front was determined as the sedimentary subfacies of the Yulou reservoir. Then, combined with electrofacies analysis, the sedimentary microfacies of the Yulou reservoir were classified (Fig. 1.4). Core observation and description can yield the most intuitive data of underground reservoirs, which, however, are limited in quantity due to the limited number of coring wells, and generally based on "one borehole's view"; therefore, it is necessary to make comprehensive analysis using sufficient core data together with regional geological setting, as well as logging and seismic data, so as to avoid errors caused by the ambiguity of geological genesis.

1.4.3 Thin Section Observation and Identification

Thin section observation and identification is also a common method used in the geological genetic analysis of oil and gas reservoirs. The features of pore space (e.g., microfractures, primary pores, and secondary pores) and diagenesis (e.g., compaction, dissolution, and cementation) can be observed in thin sections under the microscope, which can provide the most direct evidence for the geological genetic analysis of oil and gas reservoirs. Based on the observation and analysis of core thin sections, Schmidt et al. (1982) summarized the origins of secondary sandstone porosity as fracturing, shrinkage, dissolution of sedimentary material, dissolution of authigenic cement, and dissolution of authigenic replacement. Accordingly, secondary pores are divided into: intergranular pores, extra-large pores, moldic pores,

Fig. 1.3 Geological genesis analysis of oil and gas reservoirs characteristics of modern sedimentary and field outcrop data. **a** The characteristics of meandering river in Hulunbeier color belt of Inner Mongolia; **b** The characteristics of the Yarlung Zangbo River meandering river in Tibet (Namjagbarwa Peak section); **c** interlaced bedding of Mesoproterozoic Changzhougou Formation in Jixian County, Tianjin; **d** Tianjin Jixian Mesoproterozoic higher than Zhuang group ripple marks; **e** Dunhuang Happy Valley Quaternary aeolian dune-alluvial fan deposits; **f** Dunhuang, Gansu Heishanzuizi light brown red sandy mudstone abandoned channel deposition; **g** Beijing Yanqing Qianjiadian braided river sedimentary characteristics; **h** fault development characteristics of Yanchang Formation in Yanhe section of Ordos Basin, Shaanxi

Fig. 1.4 Characteristics of sedimentary core in fan delta front of Yulou oil layer in a certain area of western sag of Liaohe Basin. **a** J2, depth: 974.72–974.85 m, gray-brown glutenite, channel retention deposition; **b** J23-261, depth: 975.3–975.35 m, grayish brown siltstone, plate cross bedding; **c** J2, depth: 988.69–988.75, gray green mudstone, horizontal bedding; **d** J22-10, depth: 994.78–994.88 m, gray green mudstone, carbonated plant debris on the bedding interface; **e** J10-22, depth: 991.95–992.25 m, silty mudstone, gastropod fossils; **f** J22-10, 1049.53–1049.63 m, gray mudstone, snail fossils

intracomponent pores, and fissures opened by dissolution of authigenic replacements, by origins. I analyzed the geological genesis of the Lower Karamay Formation alluvial-fan glutenite reservoirs in an area at the Junggar Basin's northwestern margin through observing and identifying thin sections of cores obtained from different coring wells (Fig. 1.5). Thin section observation and identification, like core observation and description, is intuitive and authentic, but greatly dependent upon core samples, which are difficult to select and limited in quantity due to limited number of coring wells and research cost. In addition, the identification results are largely variable depending on the researchers' experience and competence.

1.4.4 Geophysical Prediction and Interpretation

Geophysical data includes both logging data and seismic data. The geological genetic analysis of oil and gas reservoirs in light of logging is mainly the determination and division of sedimentary facies based on electrofacies. The geological genetic analysis of oil and gas reservoirs in light of seismic includes the interpretation of fault systems based on seismic data in combination with well data, and also the prediction of reservoir sand bodies based on seismic inversion. Currently, some researchers use four-dimensional (4D) seismic data to describe the variations of oil, gas, and water during oil/gas field development. Many scholars have studied the use of geophysical data and techniques to analyze the geological genesis of oil and gas reservoirs. Qiu (2011) investigated the geologic origin of internal seismic reflection of marine strata in the Lower Yangtze region. Through the analysis of drilling data, sedimentation, diagenesis, and strata velocity, Qiu indicated that there is too strong diagenesis in the Lower Paleozoic to form any evident velocity interface and hence there are no obvious seismic reflections; while in the Upper Paleozoic Carboniferous–Permian, there are several velocity interfaces to form intense seismic reflections. When researching the reservoirs in a basin in the South China Sea, I made analysis of single-coring-well facies and seismic facies based on logging and seismic data to determine the types of sedimentary facies in the target formation for geophysical prediction and sedimentary genesis analysis of reservoirs (Fig. 1.6; Chen et al. 2012). Geophysical prediction and interpretation, as a quantitative research method, has attracted more and more attention from researchers. As the greatest advantage, it can obtain quantitative information of oil and gas reservoirs, including well site information and spatial distribution information. However, it is barely satisfactory in accuracy. On one hand, affected by the seismic processing technique and resolution, the interpretation of fault systems based on geophysical data can hardly reach the accuracy up to low-order (e.g. fifth-order) faults required for oil and gas reservoir development. On the other hand, the prediction of reservoir sand bodies based on geophysical data cannot yield a results at meter scale restricted is by acquisition conditions, depth, and other factors. Moreover, for low-permeability oil and gas reservoirs, such as the Changqing Oilfield in the Ordos Basin, seismic data are usually incomplete, but only well data are available for studies.

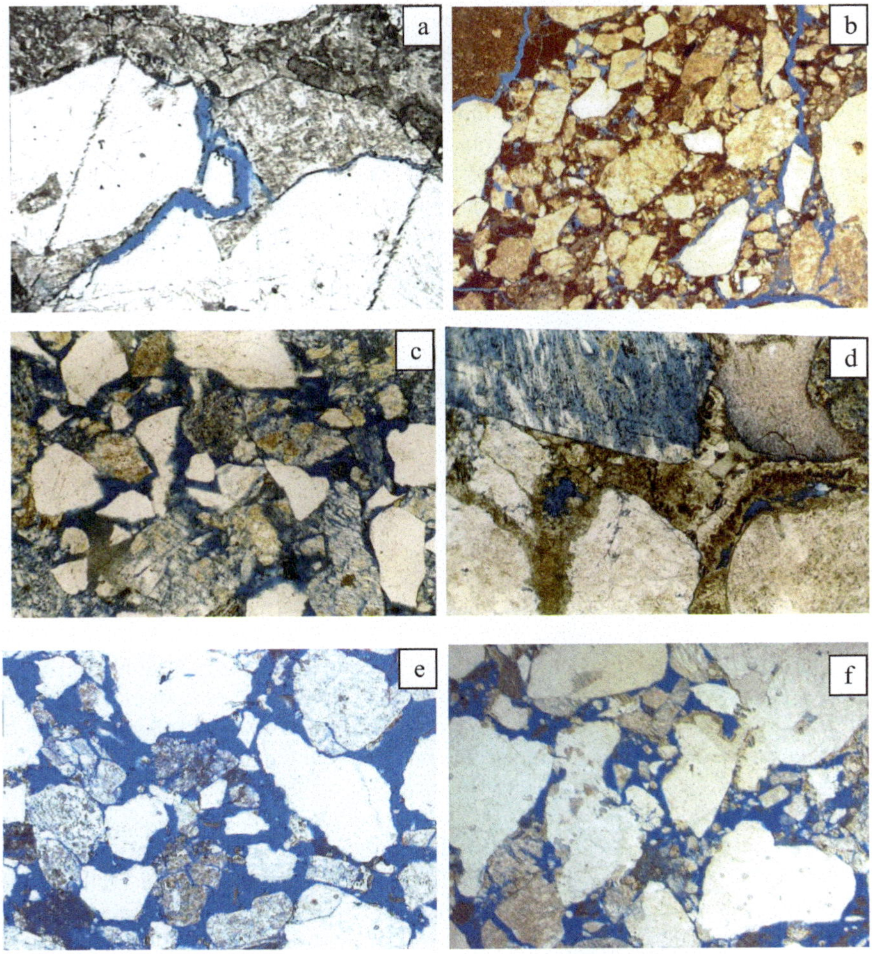

Fig. 1.5 Micro-fracture and pore structure characteristics of core slices of alluvial fan glutenite reservoir in Kexia Formation in a certain area of northwestern margin of Junggar Basin. **a** J57, glutenite, microfracture, depth: 443.19 m; **b** J6, grain-bearing coarse sandstone, micro-cracks, remaining intergranular pores, depth: 408.09 m; **c** T16, brown unequal grain sandstone, intergranular dissolved pores, intragranular dissolved pores, depth: 1017.98 m; **d** T16, glutenite, intragranular dissolved pores, dolomite dissolved pores, depth: 1098.09 m; **e** J3 well, brown gray glutenite, residual intergranular pore, depth: 416.20 m; **f** J1, gravelly coarse sandstone, residual intergranular pore, depth: 403.15 m

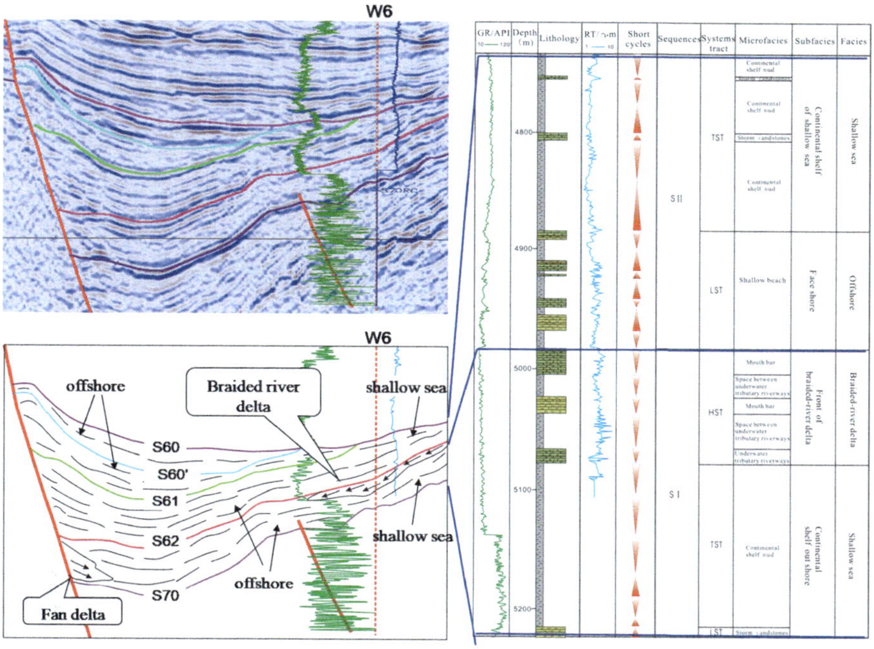

Fig. 1.6 Seismic facies and sedimentary facies demarcated section of well data and seismic date of W6 well of Lingshui Formation in shallow water area in Qiongdongnan Basin

1.4.5 Geostatistics

Geostatistics is widely used in the geological genetic analysis of oil and gas reservoirs. Zeng (2008) investigated the genesis of fractures in low-permeability sandstone reservoirs, and pointed out that the fracture angle is affected by lithology and surrounding rock (depth). Zeng made a statistical analysis on the sandstone samples from the Jilin Oilfield, China, showing that given the same lithology, the fracture strength and elastic modulus of rocks increase with the surrounding rock (depth), and the fracture angle increases obviously. Jang (2010) applied scores of multivariate statistical analyses, factor analysis (FA) and discriminant analysis (DA), to characterize the relationships between hydrochemical properties and geological origins of springs in Taiwan. Chen et al. (2014) conducted geostatistical grain-size analysis to determine the sedimentary environment of the Lower Karamay Formation alluvial-fan glutenite reservoir in an area at the Junggar Basin's northwestern margin. In the study of Chen et al., based on the statistical analysis of core grain-size data, a grain-size histogram of different rock types was plotted, indicating that the reservoir rocks are poorly sorted; moreover, the grain-size log-probability cumulative curve was drawn, revealing that the reservoir is dominated by coarse-grained sediments transported by rolling and saltation, but rare fine-grained sediments transported by

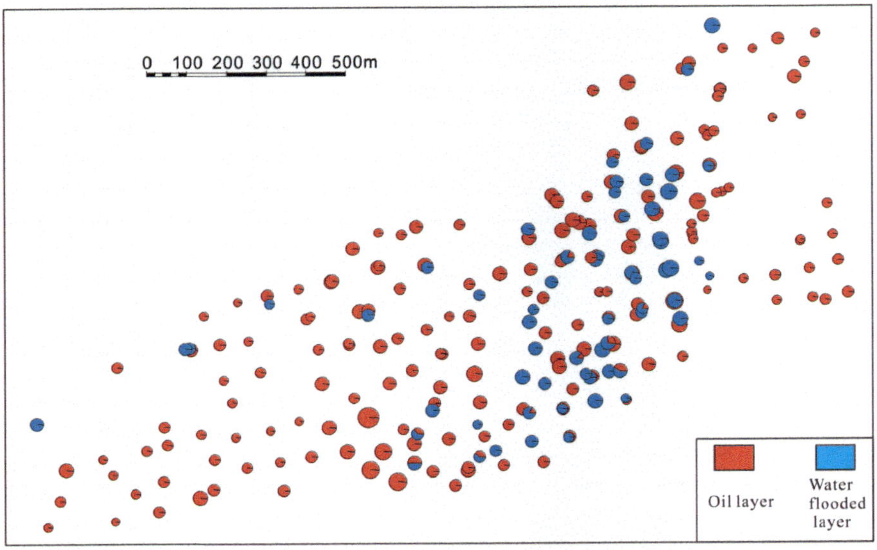

Fig. 1.7 The distribution characteristics of water flooded of single layer yI12c in the western depression of Liaohe Basin

suspension, which provides a strong evidence for the judgment of an alluvial fan sedimentary environment. In the analysis of the causes of water-flooded layers in Yulou oil layer of a certain area in the western depression of Liaohe Basin, the author first draws the plane distribution map of different single-layer water-flooded layers according to the statistical analysis of the results of fine logging interpretation (Fig. 1.7). From the diagram, the water flooded layer is mainly distributed in the central right area, and the left part of the study area to expand the experimental area of water flooded condition is much better. The cause analysis is mainly due to the invasion of edge and bottom water in the process of steam stimulation and steam flooding.

1.4.6 Analytical Tests

Analytical tests are also very important research tools for geological genetic analysis of oil and gas reservoirs. They include scanning electron microscopy (SEM), fluorescence image analysis, (constant-rate) mercury injection test, nuclear magnetic resonance (NMR) core experiment, X-ray computed tomography (X-CT) scanning experiment, nonlinear seepage test, and research on fluid inclusions in carbonate minerals (Chen et al. 2000; Huang et al. 2001; Yang et al. 2012). I applied the SEM method when analyzing the geological genesis of the Lower Karamay Formation alluvial-fan glutenite reservoir. With the SEM images, the composition of minerals can be determined, and fractures, special minerals, and diagenesis can also be observed (Fig. 1.8), which provide a solid basis for understanding the geological genesis of reservoirs. Depending on the research purpose, various test methods are widely used in the geological genetic analysis of oil and gas reservoirs. This research method provides detailed information about microscopic features of reservoirs, and can be used in researches on structures and reservoirs. However, most tests are expensive, which limits the application of this research method, especially under the climate of low oil prices where major oil companies consider endeavor to reduce cost for efficiency.

1.4.7 Physical and Numerical Simulations

Physical and numerical simulations are important to understand the geological genesis of oil and gas reservoirs. Chen et al. (2000) elaborated the method of numerical simulation of delta deposition process in a lake basin in detail. Numerical simulation of a sedimentary system uses mathematical models to describe the evolution of the sedimentary system in a basin and obtain a quantitative interpretation of relevant geological characteristics. Numerical simulation methods include flume experiment, microscale oil displacement experiment, or microscopic seepage experiment. Geological processes are mainly simulated to: (1) better understand the simulated physical processes; (2) clarify the current structure of geological elements by simulating their formation; and (3) predict the future impact of geological processes to facilitate the management of engineering and environment. Liu et al. (2014) analyzed the main factors controlling the formation and evolution of river-mouth bars in delta reservoirs, and they found that bedform slope, sediment supply, flow rate, tectonic subsidence, and relative lake water-level fluctuations have great impacts on the development of river-mouth bars (Figs. 1.9 and 1.10). In environments with flat or low-slope bedforms, the change of water depth is small, and the formed river-mouth bars are relatively easy to preserve. In such environments, the water entering the delta spreads out and the sand bodies are widely distributed, which is conducive to the preservation of river-mouth bars (Fig. 1.9). In a low-water environment, the water flow is reduced or even cut off, and the sand bodies of delta plain are mostly exposed. The hydrodynamic force of scattered small streams is not sufficient to transport coarse-grained sediments, but they may carry fine-grained sediments to form small-scale river-mouth bars with fine particles (Fig. 1.10). Flume experiment can intuitively, semiquantitatively or quantitatively analyze the genetic processes of reservoir sediments. The research method of physical and numerical simulations can intuitively observe the geological genetic processes of oil and gas reservoirs, but needs to compare the simulated process and the actual subsurface geology and appropriately simplify the basic geological genetic conditions. This process is greatly affected by the experience level of researchers. In addition, the experimental conditions and simulation software can affect the research results.

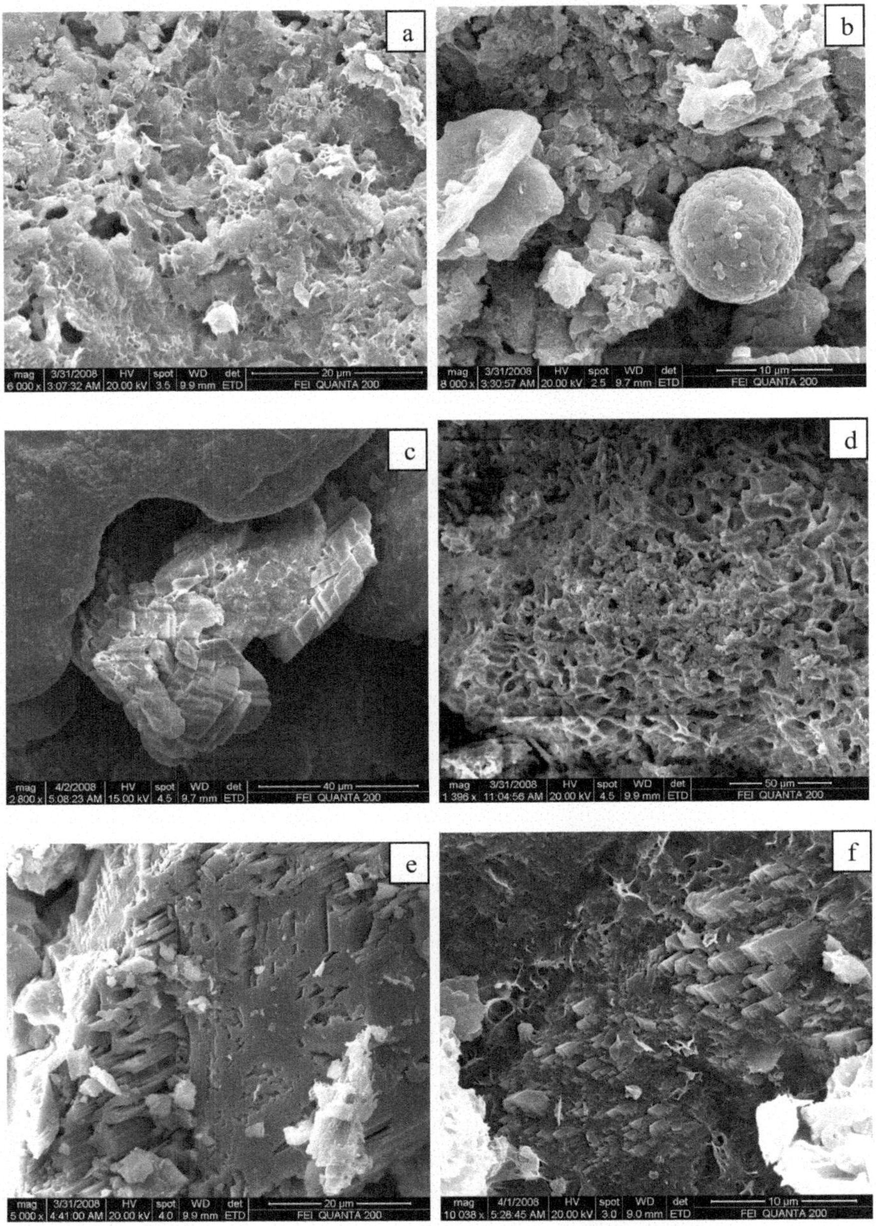

Fig. 1.8 Core scanning electron microscope characteristics of alluvial fan glutenite sedimentary reservoir in Kexia Formation in a certain area of northwestern margin of Junggar Basin. **a** J2, sandstone grain surface illite–smectite mixed layer clay, depth: 945.3 m, ×6000; **b** J2, spherical pyrite aggregate, depth: 947.2 m, ×8000; **c** J2, carbonate minerals, depth: 998.32 m, ×2800 **D** J2, biological skeleton and biological pore, depth: 961.25 m, ×1396 **E** J2, sandstone particle dissolution phenomenon, depth: 956.68 m, ×5000 F feldspar secondary enlargement grade I, depth: 979.32 m, ×10,038

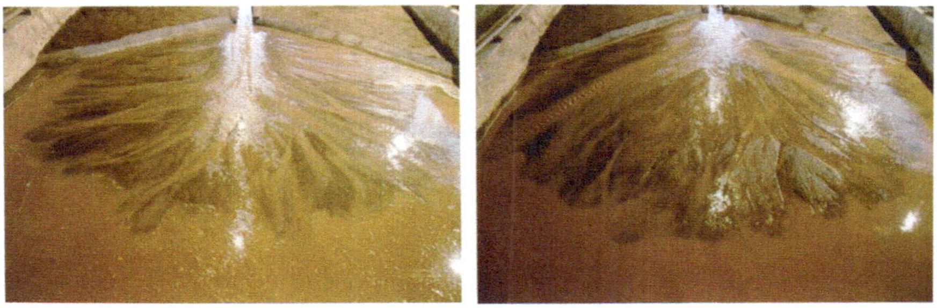

Fig. 1.9 A schematic diagram of the formation of a continuous estuary dam by a divergent water system under gentle slope (Liu et al. 2014)

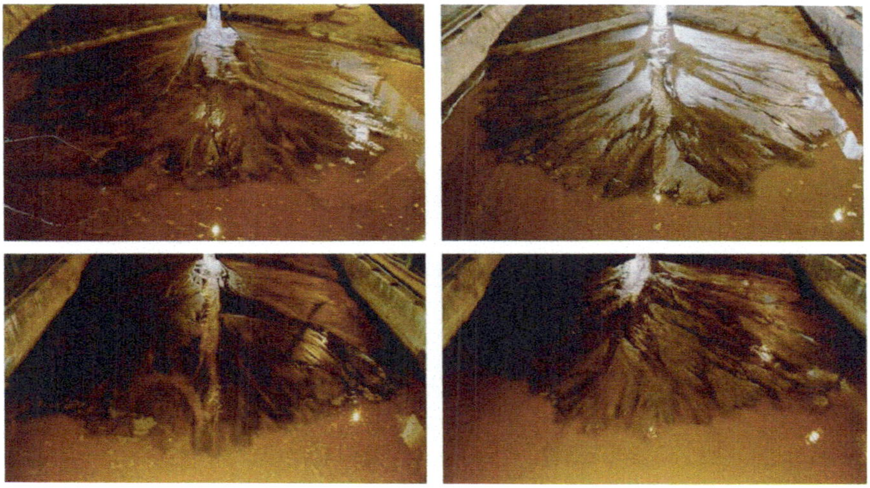

Fig. 1.10 Schematic diagram of estuary dam development in dry environment (Liu et al. 2014)

1.4.8 Reservoir Performance Monitoring and Production Performance Analysis

Reservoir dynamic monitoring and production performance analysis deduces the geological genesis characteristics of oil and gas reservoirs from the changes in various data monitored during development or in productivities of wells in different pay zones of the oil and gas reservoirs. This research method is still problematic and thus seldom used in the geological genetic analysis of oil and gas reservoirs. I used the method of reservoir dynamic monitoring to analyze the sedimentary microfacies of the Yulou reservoir in a steam-flooding test area of the Liaohe Basin's West Sag. Two flooding units are selected as examples (Fig. 1.11), where Z1 and Z2 are steam injectors, and hollow circles and solid circles represent producers without and with the steam front monitored, respectively. Fronts of steam injected into Z1 and Z2 are monitored to diagnose the connectivity of reservoirs between steam injectors and oil producers. Note that the reservoirs in the study area are loose and weakly reformed by diagenesis, so the reservoir properties are mainly controlled by sedimentary microfacies. In the flooding unit with the steam injector Z1, B1–B4 and B8 are preferential response wells, B5–B7 are subpreferential response wells, and B9–B11 are nonresponse wells. The producers with obvious responses are mainly arranged in the direction parallel to the provenance, indicating that the steam front is apparently controlled by sedimentary microfacies and mainly advances along the main stream line. B9–B11 are not affected by steam, because they are far away from Z1. Changes in main channels owing to the frequent diversions of underwater distributary channels deteriorate the reservoir connectivity. Similar situations are seen in the flooding unit with the steam injector Z2. This example proves that the reservoir dynamic monitoring method can be used to study reservoir sedimentology and determine reservoir sediment sources.

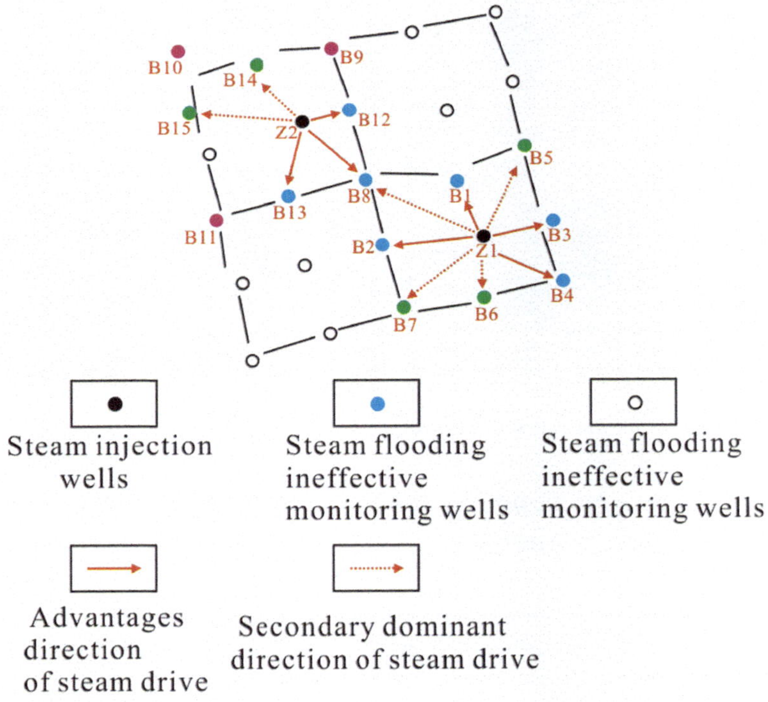

Fig. 1.11 Monitoring map of steam flooding aimed stratum of Z1 and Z2 well area of Yulou oil bearing sets in some experimental area in south end of westward slope of West depression in Liaohe Basin

1.5 Geological Genetic Analysis of Special Oil and Gas Reservoirs

At present, the geological genetic analysis of conventional oil and gas reservoirs is relatively mature. The geological genesis of conventional reservoirs is believed to be controlled by three factors: tectonism, sedimentation, and diagenesis, which play variable roles in different areas and zones. However, the geological genesis of special oil and gas reservoirs, such as low-resistivity and high-radioactivity reservoirs, remains indefinite. It is introduced specifically here.

Hui et al. (2003) analyzed the geological genesis of original low-resistivity oil reservoirs in the Dagang Oilfield, north China, and concluded that the low-energy hydrodynamic depositional conditions and secondary pores formed by diagenesis are the main geological genesis of original low-resistivity oil reservoirs. Zhou et al. (2008) presented an integrated identification method for low-resistivity hydrocarbon zones in the Nanpu Sag, the Bohai Bay Basin, China (Fig. 1.12). And they concluded that the intense variation of lithology, physical properties and formation water properties and the deep invasion of brine mud are two key factors for sharp decrease of reservoir resistivity and reduction of differences in electric characteristics between oil and water zones. As shown, Zones #33 and #39 are thin, and have equivalent porosity values and resistivity values (~ 10 Ω m). They may possibly be misjudged by simple qualitative interpretation. Analysis showed that Zone #33 exhibits finer lithology and higher content of bound water than Zone #39. According to the $R_{wa}(R_t)$—ΔGR cross plot, Zone #33 is interpreted as an oil zone, and Zone #39 as a water zone. This interpretation result was verified by the oil test results, which revealed the production of 33.8 m^3/d water and no oil from Zone #39, and of 12.7 t/d oil and 0.3 m^3/d water from Zone #33 (Fig. 1.13). Wu et al. (2015a, b, c) analyzed and evaluated the geological genesis of high-radioactivity reservoirs in the Beibu Gulf Basin, the South China Sea, based on natural gamma-ray spectral and imaging logging data, as well as core analysis data, and concluded that the high radioactivity of the reservoirs is mainly attributed to the presence of K-feldspar, which is considered, through further analysis using regional structural and sedimentary characteristics, as the result of close provenance, short-distance transport, and rapid deposition. Generally, the current researches on geological geneses of special oil and gas reservoirs mainly focus on hydrocarbon potential rather than structures and reservoir beds. The geneses of special oil and gas reservoirs are generally quite different from those of conventional oil and gas reservoirs. They should be analyzed appropriately using sufficient data beyond traditional ideas.

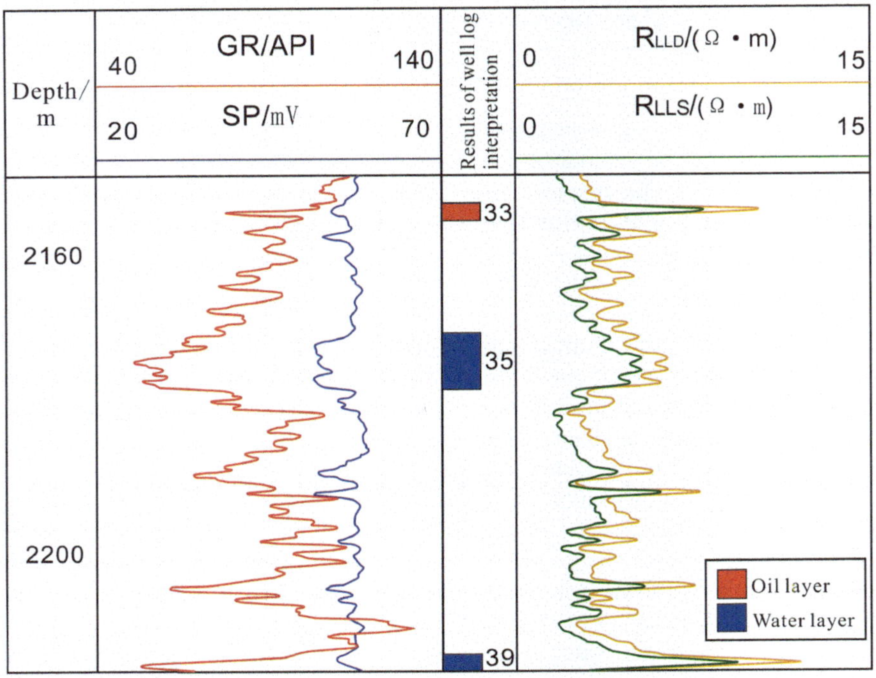

Fig. 1.12 The logging curve of ×3 well in Nangpu Sag (after Zhou et al. 2008)

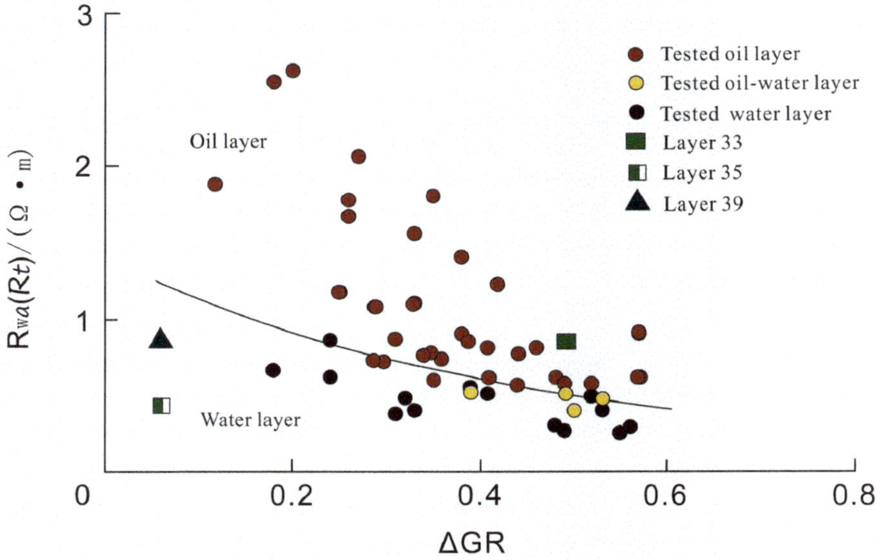

Fig. 1.13 Rwa (Rt) − ΔGR intersection chart interpretation results of ×3 well in Nanpu sag (after Zhou et al. 2008)

1.6 Problems and Prospects of Geological Genetic Analysis of Oil and Gas Reservoirs

Geological genetic analysis of oil and gas reservoirs touches all aspects of oil/gas field development. This section presents the problems and prospects thereof, which are summarized from literature survey and my practical work.

1.6.1 Problems

(1) Geological genetic analysis of oil and gas reservoirs is insufficiently concerned with field development. In the current oil/gas field development research, more attention is paid to the problem of what than why; in other words, researchers pay more attention to the characterization of structures and reservoirs but less to the analysis of their geneses.
(2) The research methods tend to be qualitative, such as investigation of field outcrops and modern sediments, core observation and description, and thin-section identification. However, physical simulation and numerical simulation methods are less used. Quantitative or semiquantitative methods need to be strengthened. It is recommended to make survey of field outcrop sections and geostatistical analysis of core and thin-section data, where possible, to realize more quantitative geological genetic analysis of oil and gas reservoirs.
(3) Geological genetic analysis and reservoir characterization are inadequately combined, which results in conflicting characterizations of oil and gas reservoirs. For example, in the quantitative classification and evaluation of reservoirs, parameters that can fully reflect the influencing factors of reservoir genesis should be selected on the basis of geological genetic analysis, and such research should be carried out with the help of quantitative calculation platforms such as SPSS. If a geological genetic analysis is missing or not done well, the rationality of parameters selected cannot be ensured, nor do the quantitative reservoir evaluation results.
(4) The precisions of interpretation and prediction of structures and reservoirs restrict the accuracy of geological genetic analysis. The accuracy of seismic data acquisition and processing limits the interpretation and prediction of microstructures and thin sand layers. It is difficult for seismic data to meet the need for adjustments in the middle and later stages of development, resulting in variations in the research results of microstructures and narrow or thin sand bodies, which affects the geological genetic analysis of oil and gas reservoirs.
(5) The geological genetic analysis of oil and gas reservoirs is incomprehensive. Due to the complexity of geological geneses and the limitations of various research data, researchers need to integrate geology, geophysics, test, and dynamic data in a comprehensive analysis to determine the main factors, providing a basis for

structure interpretation, reservoir prediction, and discrimination of oil, gas, and water zones.

(6) Many challenges exist in the geological genetic analysis of special oil and gas reservoirs. Zhu (2006) studied the microcosmic mechanism and geological origin of low-resistivity layers in the Jiyang Depression, the Bohai Bay Basin, China, and concluded that the microcosmic factors for reducing apparent resistivity of oil zone include: (a) high-salinity formation water of reservoir; (b) additional conductivity of clay minerals in the reservoir; and (c) development of micropores in the reservoir.

1.6.2 Prospects

(1) Many problems encountered in the development of oil/gas fields, such as inter-well connectivity, adjustment of injection–production well patterns, or rolling and extension, need to be solved from the perspective of geological genesis of oil and gas reservoirs, instead of merely laying emphasis on quantitative research by, for example, mathematical statistics or geological modeling.

(2) Physical and numerical simulations are used to facilitate the geological genetic analysis of oil and gas reservoirs from the quantitative perspective. Based on gravity forward modeling, Hu et al. (2009) analyzed the geological factors that may produce gravity anomalies and the gravity responses of different geological bodies in the southern Dongying Sag, the Bohai Bay Basin, China.

(3) The geological genetic analysis is further integrated into the characterization of oil and gas reservoirs to continuously improving the rationality of results. Oil/gas field development research involves many aspects, such as structures, reservoirs, and fluids, in each of which geological genetic analysis is necessary. Only an organic integration of geological genetic analysis and reservoir characterization can really improve the quality of the latter to provide a solid basis for oil/gas field development.

(4) Oil and gas reservoir characterization and geological genetic analysis complement with each other. The improvement of the former can also promote the progress of the latter. Promoting the development of geological genetic analysis of oil and gas reservoirs through the improvement of reservoir characterization will become an important research direction.

(5) The application of geological genetic analysis of oil and gas reservoirs is expanding in oil/gas field development, and it plays increasingly important roles in structure interpretation, sand body prediction, and characterization of oil, gas, and water distribution. Based on the analysis of petrology, sedimentary components, sedimentary structure, textural maturity, sedimentary cycles, and fan scale, Chen et al. (2014) determined that the alluvial fan of the Lower Karamay Formation represents a deposit transiting from humid to arid type. This provides a solid foundation for the prediction of sand body shapes, which vary greatly in different types of alluvial fans, in the study of reservoir architecture.

(6) Geological genetic analysis of special oil and gas reservoirs is a difficult but popular subject. It will certainly become an important prospect in the industry. Li et al. (2015a, b) studied the geological origin of crude oil thickening in the reservoir of the Niubao Formation in the Lunpola Basin, Tibet, China, and found through structure interpretation and core observation that faults and high steep fractures are well developed, resulting in the loss of light components of crude oil, so the oil in the reservoir is relatively thick.

The geological genetic analysis of oil and gas reservoirs can help the researchers address many problems in field development to facilitate the oil/gas exploitation and recovery. Generally, various data and methods will be combined in the geological genetic analysis for a higher quantitative level, so that specific problems in practical researches can be solved to get a deeper understanding of reservoirs, thereby providing a basis for efficient development of oil/gas fields.

1.7 Summary

The geological genetic analysis of oil and gas reservoirs is of great significance for oil/gas field development. It is conducted in three aspects: structures, reservoirs, and OGW systems, among which the geological genetic analysis of reservoirs is most extensive. The research methods for the geological genetic analysis of oil and gas reservoirs include field outcrop and modern sediment investigations, core observation and description, thin section observation and identification, geophysical prediction and interpretation, geostatistics, analytical tests, and physical and numerical simulations. These methods have their respective advantages and disadvantages, so they should be combined properly and organically in practical applications.

The geological genetic analysis of oil and gas reservoirs has six problems. (1) It is insufficiently concerned with field development. (2) The research methods tend to be qualitative. (3) Geological genetic analysis and reservoir characterization are inadequately combined. (4) The precision of reservoir characterization restricts the accuracy of geological genetic analysis. (5) The geological genetic analysis of oil and gas reservoirs is incomprehensive. (6) Many challenges exist in the geological genetic analysis of special oil and gas reservoirs.

The geological genetic analysis of oil and gas reservoirs will develop in six prospects. (1) Many problems encountered in the development of oil/gas fields need to be solved from the perspective of geological genesis of oil and gas reservoirs. (2) Physical and numerical simulations are used to facilitate the geological genetic analysis of oil and gas reservoirs from the quantitative perspective. (3) The geological genetic analysis is further integrated into the characterization of oil and gas reservoirs to continuously improving the rationality of results. (4) Oil and gas reservoir characterization and geological genetic analysis complement with each other; the improvement of the former can also promote the progress of the latter. (5) The application of geological genetic analysis of oil and gas reservoirs is expanding in oil/gas field development. (6) Geological genetic analysis of special oil and gas reservoirs will certainly become an important prospect in the industry.

References

Agha MA, Hart GF, et al. Minerlogy of Egyptian bentonitic clays II: geologic origin. Clays Clay Min. 2013;61(6):551–65.

Chen H. Quantitative estimation of inter layers heterogeneity characteristics of volcanic reservoir: taking one member of Yingcheng formation in Xudong area for an example. J China Univ Min 8L Technol. 2012;28(1):58–67.

Chen F. The geological genesis of calcareous mudstone and its petroleum geological significance of Nantun Formation in Beier Sag. J Northeast Petrol Univ. 2015;39(2):42–50, 93.

Chen L, Wang J, Li Y, et al. Oil and gas reservoir research technology. Beijing: Petroleum Industry Press; 2000.

Chen H, Hu Y, Ran Q, et al. Mode characteristics of volcanic eruption of the member 1 of Yingcheng Formation in Xudong Area in Songliao Basin. J Southwest Petrol Univ (Sci Technol Edn). 2012;34(1):41–48.

Chen H, Zhao Y, Gao X, et al. High-resolution sequence stratigraphy and its application in the fine-scale stratigraphic correlation of the Yulou Reservoir in the West Depression of the Songliao Basin. J Stratigr. 2014;38(3):317–23.

Chen H, Hu Y, Shi C. Single factor characterization and quantitative evaluation of volcanic rock reservoirs in Songliao basin. Beijing: Petroleum Industry Press; 2016a.

Chen H, Shi C, Chen C. Problems with the fine description of reservoirs. Pet Geol Exp. 2016b;38(5):569–76.

Chen H, Shi C, Hu H, et al. Reflection and countermeasure of fine reservoir description in low oil prices. Geol Sci Technol Inform. 2017;36(5):85–91.

Chen H, Hu H, Wu H, et al. Geological genetic analysis of oil and gas reservoir development. Chin J Geol. 2019a;54(1):192–212.

Chen HL, Tang XY. translated. Secondary reservoir pores in sandstone diagenesis. Beijing: Petroleum Industry Press, 1982.

Chen P, Qibo W, Lina G, et al. Multi-parameters quantitative evaluation for carbonate reservoir based on geological genesis. J Southwest Petrol Univ (Sci Technol Edn). 2019b;41(4):55–64.

Cheng-Shin J. Applying scores of multivariate statistical analyses to characterize relationships between hydrochemical properties and geological origins of springs in Taiwan. J Geochem Explor. 2010;105:11–8.

Feng Z. Sedimentary petrology. Beijing: Petroleum Industry Press; 1994.

Furukawa K, Uno K, Shinmura T, et al. Origin and mode of emplacement of lithic-rich breccias at Aso Volcano, Japan: Geological, paleomagnetic, and petrological reconstruction. J Volcanol Geoth Res. 2014;276:22–31.

Gao L, Ming J, Yan T, et al. Application of seismic attribute analysis technique to the identification of mudstone interbeds. Lithol Reserv. 2013a;25(4):101–5.

Gao Y, Sun W, Xin Z. Microscopic geologic origin for ultra-low permeability reservoir of Yanchang Groups, Ordos Basin. J Jilin Univ (Earth Sci Edn). 2013b;43(1):13–9.

Hogan JF, Phillips FM, Mills SK, et al. Geologic origins of salinization in a semi-arid river: the role of sedimentary Basin brines. Geology. 2007;35(12):1063–1066.

Hu J, Sui Z, Liu C. The cause of gravity anomaly in the south of dongying sag. Petrol Geol Recov Eff. 2009;16(2):39–42.

Huang Y, Yu D. Micro-seepage experiment mechanics and its application. Beijing: Petroleum Industry Press; 2001.

Huang D, Hua W, Chen L, et al. The geological origin of carbonate reservoirs in southern China—take the Qixia Formation of lower Permian in Western Sichuan Province as an example. Nat Gas Geosci. 2012;32(11):22–6.
Hui X, Wu X, Xie Q, et al. The geologic genesis of original low resistivity oil reservoirs in Dagang oil field. Progr Explor. 2003;26(4):329–32.
Jang C-S. Applying scores of multivariate statistical analyses to characterize relationships between hydrochemical properties and geological origins of springs in Taiwan. J Geochem Explor. 2010;105:11–18.
Jiajun Z. Geologic origin and microcosmic mechanism of low-resistivity layers in Jiyang Depression. Acta Petrolei Sinica. 2006;27(6):43–6.
Kotarba MJ, Nagao K, Karnkowski PH. Origin of gaseous hydrocarbons, noble gases, carbon dioxide and nitrogen in Carboniferous and Permian strata of the distal part of the Polish Basin: geological and isotopic approach. Chem Geol. 2014;383:164–79.
Li H, Liu S, Wei X, et al. Method exploration of implicit logging geological information research. Prog Geophys. 2015a;30(1):195–202.
Li W, Ren J, Liu Y, et al. Development and formation mechanism of Cenozoic tectonics in the East Area of Liaodong Bay Depression. Geol Sci Technol Inform. 2015b;34(6):58–63.
Liu Y, Chen Q, Hu K, et al. Comparison of the Bohai Bay Basin and Subei-South Yellow Sea Basin in the Structural Characteristics and Forming Mechanism. Geotecton Metallog. 2014;38(1):38–51.
Lobach-Zhuchenko SB, Rollinson HR, Chekulaev VP, et al. The Archaean sanukitoid series of the Baltic Shield: geological setting, geochemical characteristics and implications for their origin. Lithos. 2005;79:107–28.
Malone MJ, Claypool G, Martin JB, et al. Variable methane fuxes in shallow marine systems over geologic time: the composition and origin of pore waters and authigenic carbonates on the New Jersey Shelf. Marine. 2002;189:175–96.
Qiu X. Geologic origin of the internal seismic reflection of marine strata in the Lower Yangtze region. Oil Gas Geol. 2011;32(3):397–403.
Qiu Y, Chen Z. Reservoir description. Beijing: Petroleum Industry Press; 1996.
Sarmah MK, Borthakur A, Dutta A. Interfacial and thermal characterization of asphaltenes separated from crude oils having different geological origins. Pet Sci Technol. 2010;28:1068–77.
Wu J, Hu X, Liang Y, et al. Geological formation analysis and evaluation of high radioactive reservoir in Beibu Gulf basin. Spec Oil Gas Reserv. 2015a;22(1):79–83.
Wu L, Chen Q. Structural characteristics and evolution of basement faults in Gaoyou Sag, Subei Nasin. Nat Gas Geosci. 2015b;26(4):689–99.
Wu Y, Chang L, Chen W, et al. Application of fracture characterization and modeling in Dina 2 Gas Field. Fault-Block Oil & Gas Field. 2015c;22(1):78–81.
Xu H, Lu J, Liang J. Seismic identification and geological origin of gas hydrate in near seafloor sediments in the eastern part of the Pearl River Mouth Basin. Earth Sci Front. 2017;24(4):057–65.
Yang Z, Guo H, Liu X, et al. Characteristic experimental technology of ultra-low permeability reservoir. Beijing: Petroleum Industry Press; 2012.
Yuce G, Italiano F, Alessandro WD, et al. Origin and interactions of fluids circulating over the Amik Basin (Hatay, Turkey) and relationships with the hydrologic, geologic and tectonic settings. Chem Geol. 2014;388:23–39.
Zeng L. Formation and distribution of cracks in low permeability sandstone reservoirs. Beijing: Petroleum Industry Press; 2008.
Zeng L, Ke S, Liu Y. Research method of low permeability oil and gas reservoir fracture. Beijing: Petroleum Industry Press; 2010.
Zhang Y, Qin Q. Analysis of sandstone reservoir micro-pore structure fractal based on geological origin. Nat Sci J Xiangtan Univ. 2018;40(3):95–9.
Zhang F, Cao G, Li Y, et al. Used of seismic properties analysising technology in researches of volcanic cracks in Xudong Area in the North of Songliao Basin. J Daqing Petrol Inst. 2007;31(2):12–4.

Zhou F, Si Z, Ma Y, et al. Integrated identification method for low-resistivity hydrocarbon layers in Nanpu Sag. Pet Explor Dev. 2008;35(6):680–4.

Zhu J. The geological cause analysis of high oil saturation in low porosity and low permeability reservoirs—taking Shengli Oilfield in Dongying Depression. J Oil Gas Technol. 2008a;30(3):64–7.

Zhu X. Sedimentary petrology, 4th edn. Beijing: Petroleum Industry Press; 2008b.

Chapter 2
Geological Genetic Analysis of Structures

2.1 Progress in Geological Genetic Analysis of Structures in Field Development

Geological genetic analysis of structures focuses on extremely different aspects from stage to stage. At the stage of exploration and appraisal, geological genetic analysis of structures mainly deals with the evolution history of structure, development history of the fault system, phases and interaction of faults, in situ stress orientation, and development and distribution of the third- and higher-order faults. At the stage of development, the geological genetic analysis of structures treats the development of third- and lower-order faults and the development of microstructures, aiming to achieve enhanced oil and gas recovery and remaining oil characterization. However, during field development, especially at the middle–late stage, most studies pursue quantification along with the vigorous growth and application of computer technologies and various mathematical and physical methods. In contrast, efforts concerning structures from the perspective of geological genesis are weakened and attenuated. The tendency toward an "indoor geologic study" has serious consequences in practice. For example, in CNPC oilfields, either in eastern or in western China, the structure interpretation result for a horizon in the same block changes every year, sometimes remarkably, and even at the stage of development, the interpretation of the third- and higher-order faults may be changed. Such a change is attributed certainly to the availability of more data and experience but fundamentally to an insufficient understanding of the geological genesis of structures. It is believed to be avoidable or mitigatory to a great extent given that the structural evolution phases, in situ stress orientation, and rock types/features of target horizons in the study area are clarified. Fine and accurate structure interpretation is a hard nut to crack in fields, which seriously impedes the efficient field development and the implementation of techniques for enhanced oil/gas recovery. In this section, the geological genetic analysis of structures in field development is discussed from the aspects of literature survey, research content and methods, and problems and prospects.

© Petroleum Industry Press 2024

H. Chen, *Geological Genetic Analysis and Application of Oil and Gas Reservoirs*,
https://doi.org/10.1007/978-981-97-7628-3_2

2.1.1 Literature Survey

Geological genetic analysis of structures is incorporated in works on almost all types of minerals, including but not limited to oil and gas. An overall survey on the literature involving the geological genetic analysis of structures with respect to various solid minerals will be referential for the application of this technique in oil/gas field development.

Hofmann et al. (2003) investigated the tectonic origin of the iron ore belt in the Zimbabwe craton and its geologic significance to the greenstone belt. Li et al. (2004a, b) discussed the eustasies at the Great Australian Bight and the tectonic origin of the Neogene unconformity. James et al. (2006) conducted a study of the Pliocene Roe Plains in the Great Australian Bight and the genesis of the calcarenite from the perspectives of sedimentation and tectonism, demonstrated that the calcarenite possibly extends from the position of basin inversion to the passive continental margin. McMechan (2007) reported the nature, origin, and tectonic significance of the anomalous transverse structures in the southeastern Skeena Fold Belt, British Columbia. Millán et al. (2008) investigated the ferromanganese crusts found in the Betic Cordilleras, southeastern Spain, and considered the crust formation as the product of tectonism and sedimentation in accordance with geochemical and mineralogical data. Sainz et al. (2009) probed the tectonic origin of the present-day topography of the Iberian Peninsula. Charvet et al. (2010) focused on the Lower Paleozoic belt of South China and the genesis of an intracontinental orogen through a case study of Wuyishan and surrounding areas based on the data of structures formed before the Devonian unconformity. Yang et al. (2013) identified the structural genesis of the Eunsan and Moisan low-sulfidation epithermal Au–Ag deposits in the Seongsan district, Southwest Korea. Ibouh et al. (2014) examined the tectonic–karstic origin of the alleged "impact crater" in Lake Isli, Imilchil District, High Atlas, Morocco, and concluded that no geologic evidence supported a meteoritic origin. Botros (2015) probed the role of granite emplacement and structural setting in the genesis of gold mineralization in Egypt. Saalmann et al. (2016) presented the activation of an oblique ramp during convergence in the East African Orogen by investigating the age, tectonic evolution, and origin of the Aswa Shear Zone in Uganda. The Aswa Shear Zone is an NW–SE striking, NE-dipping steep structure of over 1000 km long and up to 11 km wide in East Africa. The mylonitic shear zone shows multiphase brittle reactivation. Cooperdock et al. (2018) used the geochemical signatures of abyssal origin preserved in an HP/LT subduction complex and the analysis of whole rock trace element, major element geochemistry, and stable isotopes (δD and δ18O) to deduce the tectonic origin of serpentinites in Syros, Greece.

Several Chinese scholars have worked on the geological genesis of structures. Liu et al. (1998a, b) analyzed the mechanisms and tectonic models of fault block reservoirs from the perspectives of fault sealing and fault block group patterns, with a case study on the Bohai Bay Basin, a faulted basin in East China. Zhang et al. (1999)

presented four controlling factors, namely, lithology, fault, depth, and ancient weathering, for structural fractures in metamorphic reservoirs. Zeng et al. (2001) summarized the research methods for fractures in low-permeability reservoirs: geological analysis, conventional-logging identification and evaluation, imaging-logging identification and evaluation, seismic fracture detection, structural fracture prediction, and fracture analysis through petroleum reservoir engineering. Yuan et al. (2004) investigated fractured oil reservoirs using the data of outcrop, core, logging, production performance, and stress field. Xie (2006) studied the genesis of Cenozoic structures in the South China Sea through numerical simulations. Li et al. (2009a, b) discussed the features of microstructures and their control on oil and water distribution with a case study of the Gaoji Oilfield in Jiangsu, China. Zhang et al. (2012) combined high-precision GPS monitoring data acquired in 2011–2008 in the Weihe Basin with regional structural features to build a finite-element kinetic model of the basin, by which they examined the present-day crust stress field and the internal relationship between the tectonic stress field and ground fractures in the basin. Moreover, they quantitatively revealed the internal kinetic relation between the regional tectonic stress field and ground fractures in the basin and the fundamental cause of unbalanced distribution of ground fractures in the eastern and western basin-based on space geodesy for the first time. Liu et al. (2014a, b) performed a comparative study of structural features and genesis in the Bohai Bay and the North Jiangsu–South Yellow Sea Basin. Zhang et al. (2015) analyzed the structural features and genesis of the Tongcheng fault in the Jinhu Sag with the aid of 3D high-precision seismic data. Hu et al. (2016) reported the structural evolution and genesis of the Chepaizi salient, the Junggar Basin, by reproducing the structural evolution process after physical simulation experiments. Dong et al. (2017) discussed the stress mechanism of the Qudi faulted nosing structure in the Jiyang Depression and its control on structures and hydrocarbons. Wang et al. (2018) probed into the features and genesis of low-relief structures on the Yishaan slope, the Ordos Basin and revealed that low-relief structures occur widely on the Yishaan slope and the tectonic stress field dominated the oriented extension, array folding, large-scale occurrence, regional distribution, and inherited evolution of low-relief structures in the region. As indicated by the comparison (Table 2.1), the geological genetic analysis of structures abroad covers various aspects and involves diverse approaches/techniques, with high-level geological modeling. In China, the analysis focuses more on fractures and demonstrates high-level physical simulation; especially, valuable progress has been made in the geological genetic analysis of microstructures. It is recommended that the advanced experience abroad in geological genetic analysis of structures be incorporated into field development studies in China.

Table 2.1 Comparison of geological genetic analysis of structures in China and abroad (Liu et al.1998a, b; Zhang et al. 1999; Corfield et al. 2001; Hofmann et al. 2003; Li et al. 2004a, b; Yuan et al. 2004; James et al. 2006; Xie et al. 2006; Margot 2007; Jimenez-Millan et al. 2008; Casas-Sainz et al. 2009; Li et al. 2009a; Charvet et al. 2010; Zeng et al. 2010; Zhang et al. 2012; Yang et al. 2013; Ibouh et al. 2014; Yin et al. 2014b; Botros 2015; Zhang et al. 2015; Saalmann et al. 2016; Hu et al. 2016; Dong et al. 2017; Cooperdock et al. 2018; Wang et al. 2018)

	Advantages	Disadvantages
Abroad	(1) Geological genetic analysis of structures is extensively conducted with great progress in light of solid minerals (2) The research covers multiple aspects of various structures, such as faults and fractures (3) A number of research methods are available, including outcrop, petrology, fluid inclusion analysis, isotope analysis, geochemistry, well-seismic structure interpretation, and basin numerical simulations (4) Geological modeling is successfully used in geological genetic analysis of structures (5) The research targets reservoirs of various genetic types, including clastic reservoir, carbonate reservoir, volcanic reservoir, and metamorphic reservoir (6) Geological genesis and spatial distribution of minerals are delineated by geological genetic analysis of structures (7) Geological genetic analysis of structures is combined with analysis of sedimentary facies and diagenesis to investigate geological genesis of reservoirs, thereby supporting the solid mineral prediction and petroleum exploration and development (8) Geological genetic analysis of structures is attractive worldwide, such as America, Asia, Europe, Africa, and Oceania	(1) Geological genetic analysis of structures is rarely made for petroleum exploration and development (2) Geological genetic analysis of structures is often macroscopic but seldom microscopic (e.g., fractures) (3) Well logging data are insufficiently used (4) Geological genesis of microstructures, especially its role in remaining oil characterization, is not considered adequately
China	(1) Geological genetic analysis of structures is made for petroleum exploration and development in addition to solid minerals (2) The research focuses on both structures (macroscopic) and fractures (microscopic) (3) The research methods include basic geology, well-seismic interpretation, and physical simulation of tectonic origin (4) Geological genetic analysis of structures is successfully and fruitfully done from the microscopic perspective through core observation and description, thin-section observation, and physical simulation (5) Great attention is paid to the geological genesis of low-relief structures, particularly minor faults and microstructures, and remaining oil characterization based on geological genesis analysis of microstructures has achieved primary effects in practice (6) Genetic models of structures are especially collected to support reservoir description and prediction	(1) Geological genetic analysis of structures is not adequately integrated with sedimentology and diagenesis (2) Many problems exist in geological modeling, particularly the geological modeling of fractures (3) Methods for geological genetic analysis of structures are more qualitative (e.g., geology, geophysics, and physical simulation) than quantitative (4) Geological genetic analysis of structures involves clastic rocks but rarely carbonate and volcanic rocks

2.1.2 Geological Genetic Analysis of Structures in Field Development

Geological genetic analysis of structures in field development roughly involves eight aspects.

(1) Structural evolution history and in situ stress field are investigated to determine the geological genesis of structures such as fault systems, providing the basis for accurate description of the grouping and development of faults. This is a focus in petroleum exploration, but not be emphasized in field development. Note that the results of geological genetic analysis of structures obtained in the stage of petroleum exploration should be checked in the during field development stage to validate the accuracy of the division of structural evolution stages, grouping of third- and higher-order faults and in situ stress orientation for subsequent activities.
(2) Fine structure interpretation and fault grouping are performed using well and seismic data to delineate structures and analyze their geological genesis (Figs. 2.1 and 2.2). Genetic models of structures are built, and their controls on hydrocarbon accumulation are analyzed. Corver et al. (2009) classified the rifted sedimentary basins of the Pannonian Basin System according to the structural genesis, evolutionary history, and hydrocarbon maturation zones through several steps: (a) to identify standard evolutionary patterns related to geodynamic drivers, (b) to group the basins with similar tectono-stratigraphic evolution and correlate these with petroleum system development, (c) to relate basin trap types to basin history, and (d) to calculate characteristic hydrocarbon maturation zones by thermal modeling. Li et al. (2013) analyzed the genetic types of Mesozoic low-relief uplifted structures in the Ordos Basin and their control on hydrocarbon distribution and concluded that many low-relief uplifted structures extend regularly on the east and west slopes of the Tianhuan Depression in the Ordos Basin, of which large low-relief nose-like uplifts extending in rows in a nearly east–west direction in the Yanchang Formation were dominated by basement paleohighs, and local uplifts in the Yanchang and Yan'an formations were jointly controlled by fault-related folds, differential compaction, and compound genesis.

(3) Methods for geological genetic analysis of structures, including outcrop observation, logging interpretation, seismic interpretation, lab test, physical simulations, and numerical simulations, are improved to the accuracy/precision required for field development. Izard et al. (2009) used fault population analysis methods, fluid inclusion research, and stable isotope data to determine the genesis and evolution of the structurally controlled vein mineralization (Sb-Hg) in the Escarlati deposit, León, Spain.
(4) Geological genetic analysis of reservoirs is conducted using outcrop, core, thin section, logging, seismic, test, dynamic monitoring, and production performance data to determine major controls on structural genesis. The geological

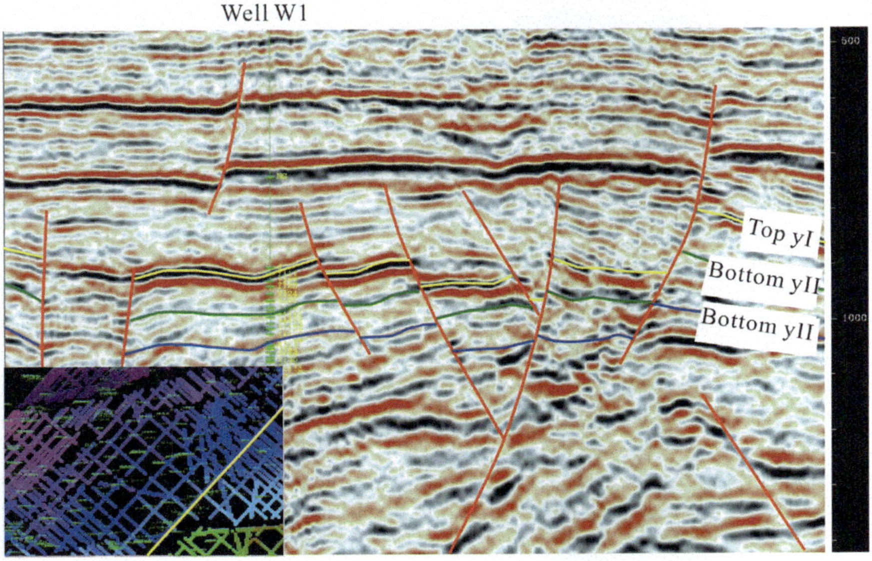

Fig. 2.1 The characteristics of a steam flooding test area in the western sag of Liaohe Basin combined with the well seismic and fracture interpretation profile

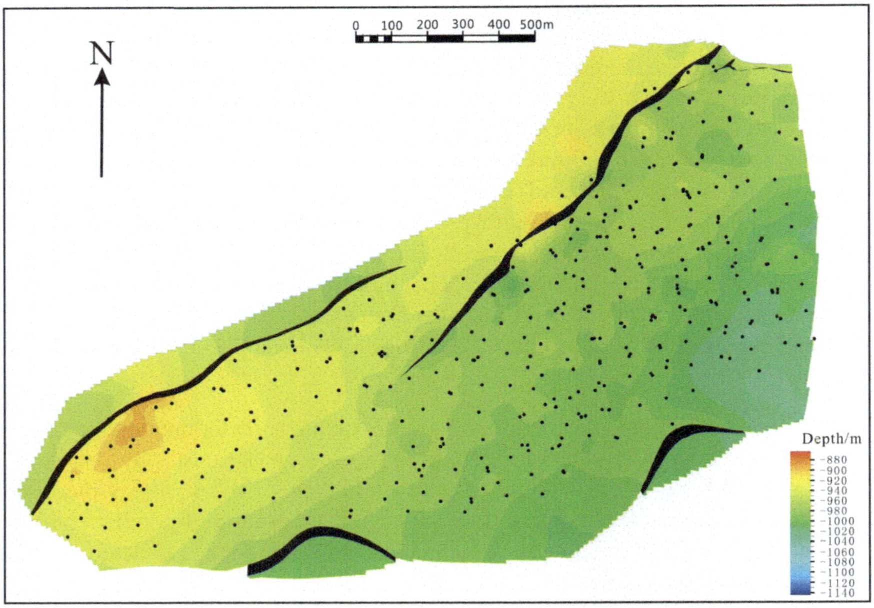

Fig. 2.2 Structure map of YI24a single layer in Yulou oil bearing sets in Western Depression of Liaohe Basin

genesis of structures is generally controlled by multiple factors, which are identified comprehensively using various data. Particularly, in addition to the data acquired at the stages of exploration and appraisal, more data will be collected at the stage of field development with consideration to both cost and benefit.

(5) Geological genetic analysis of structures is combined with studies on sedimentary facies and diagenesis to detect major controls on hydrocarbon accumulation and support development activities.

(6) Fault-sealing analysis, reservoir fracture characterization, reservoir architecture analysis, and remaining oil characterization are performed on the basis of geological genetic analysis of structures. Rothery et al. (2001) re-evaluated the tectonic origin of the Broken Hill lodes in a high-grade shear zone and concluded of structural support for ore formation.

(7) Fracture characterization and modeling are conducted, including analysis of the genetic mechanism of fractures from the perspective of geology, fracture origin classification, spatial delineation of fractures, and interwell fracture prediction. Wang et al. (2017) approached the genesis and evolution of structural fractures in the Keshen-2 gas field, the Tarim Basin. There are three groups of structural fractures extending in nearly E–W, nearly N–S, and NW–SE directions in the Keshen-2 gas field. The nearly E-W structural fractures were dominated by early regional extension, late anticlinal bend-related arching, anomalous high fluid pressure, stress transition during structural inversion, and reverse thrusting. The nearly N–S structural fractures were dominated by tectonic compression in a nearly N–S direction. The NW–SE structural fractures were generated by the maximum horizontal compressive stress in a NNW–SSE or nearly N–S direction.

(8) The geological genesis of microstructures is analyzed. Microstructures, also entitled sedimentary or reservoir microstructures, refer to local low-relief structures and inconspicuous minor faults at the reservoir top under the overall structural setting of the field (Zhang et al. 2011). Remaining oil characterization is difficult during field development, especially at the middle and late stages, when remaining oil is highly scattered across the field and relatively concentrated locally (Han 2010). Microstructures play a vital role in controlling the remaining oil distribution. Thus, microstructures, particularly their geological genesis, have been taken seriously as a focus during geological genetic analysis of structures.

2.1.3 Techniques for Geological Genetic Analysis of Structures

There are diverse techniques for geological genetic analysis of structures depending on fundamental data, research targets, and researchers' disciplines. Here, the techniques are classified into 10 categories.

(1) Basic geologic analysis

Basic geologic analysis focuses on basin structural evolution to understand tectonic stress field, structural features about the target horizon, and especially the fault system. It is a commonly used method in petroleum exploration and should also be considered in field development because inaccurate information about regional structural features may lead to a misjudgment of small-scale structures in field development. In a study of Mesozoic tectonic styles and their dynamic mechanisms in the East China Sea Shelf Basin, Yang et al. (2017) classified tectonic styles into extensional structures, compressional structures, strike-slip structures, inversion structures, and diapir structures, which are further divided into 12 tectonic assemblages. Basic geological analysis is the most fundamental technique for understanding the geological genesis of structures, and it helps acquire structural features from a macroscopic viewpoint. However, this technique is more qualitative than quantitative. Many researchers have mitigated this weakness through tectonic stress field analysis based on rock mechanical experiments, which need to be improved in selection of representative data, comparison of experimental results with actual subsurface geologic conditions, and other aspects.

(2) Outcrop observation and description

Outcrop observation and description is a direct and effective technique for geological genetic analysis of structures. It gathers information on the fault, such as property, size, episode, and strike and helps in the analysis of the size and geometry of folds; the size of vugs and fractures; and the fillings, occurrence, and geometry of fractures. Outcrop candidates should be representative and can be correlated with subsurface geology. A specific number of samples are necessary to avoid the fallacy of composition. If necessary, outcrops should be sampled for the study of lithology, rock mechanics, and microscopic rock texture. Ge et al. (2015) investigated the features and dynamic mechanism of the southern end of the Muping-Jimo fault in the western coastal region of Qingdao, Shandong Province, China, based on a detailed field geologic survey and found that the structure of the fault along the western coast of Qingdao mainly consists of numerous shear joints, as well as faults and dikes. I made a satisfactory geological genetic analysis of structures by observation and description of outcrops in Wulong County in Chongqing, Yanqing District in Beijing and Jizhou District in Tianjin (Fig. 2.3). Through detailed outcrop observation and description, the researchers can gain a direct insight into the structural features of fault and fold, which include fault size, fracture occurrence, and fracture/pore fillings. In addition to macroscopic features, centimeter-scale information can be obtained through local microstructure observation of outcrops. This intuitive and accurate technique has been widely applied to the geological genetic analysis of structures. It helps in deep understanding of the genetic mechanisms and models of structures for subsequent fine structure interpretation. However, outcrop conditions are somewhat different from subsurface geology of an oil and gas field. Such discrepancies should be addressed before the deliverables of outcrop observation are integrated into the production.

(3) Core observation and description

Fig. 2.3 Geological genesis analysis of field outcrop structure. **a** Field outcrop characteristics of the fault system in Pengshui County, Chongqing, Wujiang section; **b** field outcrop characteristics of fault system in Wulong County, Chongqing; **c** outcrop characteristics of Beijing Yanqing fold outcrop characteristics; **d** Tianjin Jixian Great Wall fracture system; **e** outcrop characteristics of Tianjin Jixian fault system; **f** field outcrop characteristics of Tianjin Jixian fault system, fracture-cavity in limestone of Tieling Formation

Core observation and description is also a common and intuitive technique used in geological genetic analysis of structures during field development with abundant core data and even sealed cores in some major blocks. Through core observation, the researchers can understand the geological genesis of structures and even the tectonic impact of petroleum recovery on reservoirs, such as dynamic fractures generated during production. The author used core observation and description to examine the fractures in the volcanic reservoirs in the first member of the Yingcheng Formation (Ying-1 Member) in the Xudong area, the Songliao Basin, and glutenite reservoirs in the Lower Karamay Formation at the northwest margin of the Junggar Basin (Fig. 2.4). Well core observation and description yield visualized images of fracture size, geometry, and density. If there are sufficient core data, fracture density can be quantitatively estimated for semiquantitative fracture delineation. This technique is intuitive and accurate, with easy availability of data. However, it is feasible for the study of fractures but not for large faults. It can derive proper conclusions only in conjunction with analysis of the regional tectonic setting.

(4) Thin-section observation

Thin-section observation is another intuitive technique for geological genetic analysis of structures. The episode, genesis, size, and geometry of microfractures can be

Fig. 2.4 Core observation of micro-fracture characteristics. Fracture characteristics of volcanic reservoir in the first member of Yingcheng Formation in Xudong area of **a** Songliao Basin, X21 well, depth: 3656.24–3656.58 m; **b** Fracture characteristics of volcanic reservoirs in the first member of Yingcheng Formation in Xudong area of Songliao Basin, X21 well, depth: 3658.72–3659.15 m; the fracture characteristics of glutenite reservoir in Kexia Formation of a certain area in the northwest margin of **c** Junggar Basin, J1 well, depth: 389.2–389.5 m; **d** fracture characteristics of glutenite reservoir in Kexia Formation in a certain area of the northwestern margin of Junggar Basin, Well J1, depth: 393.48–393.70 m

observed on thin sections under a microscope using plane-polarized light or cross-polarized light. Similar fractures can be inspected through thin-section observation and core observation, except for the smaller scale of thin-section fractures. The author used thin-section observation to investigate the fractures in the volcanic reservoirs in the Ying-1 Member in the Xudong area, the Songliao Basin, and clastic Yulou oil reservoir in the steam-drive test area, the West Sag of the Liaohe Basin (Fig. 2.5) and obtained important information on fractures, such as genesis, extension, the state of filling, and density. Figure 2.5a shows dissolution fractures; Fig. 2.5b shows burst and structural fractures; Fig. 2.5c, d show structural fractures. This technique can yield intuitive results with low cost. Similar to core observation and description, it is feasible for the study of fractures but not for faults.

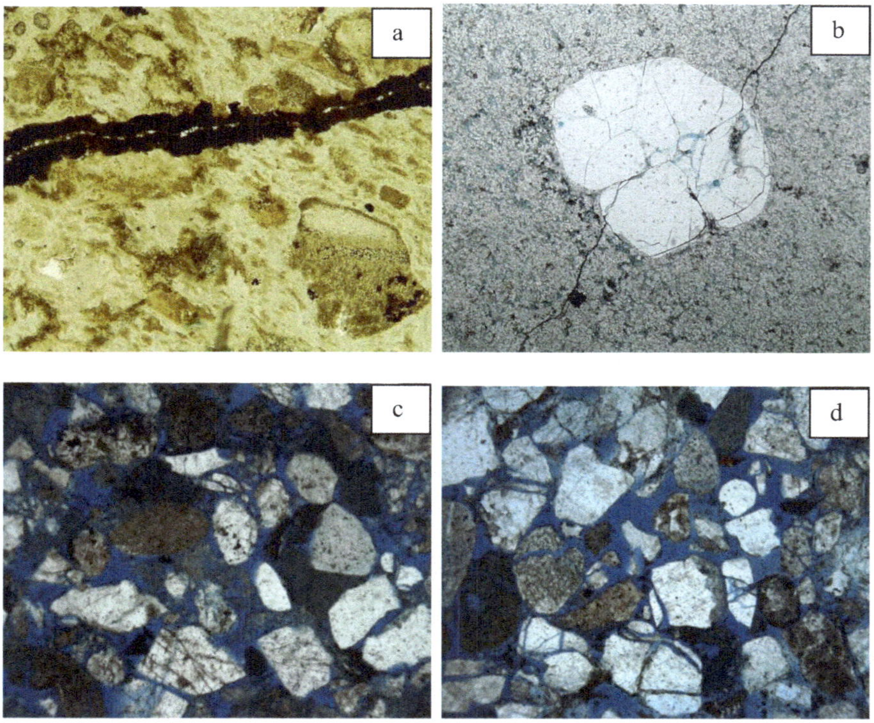

Fig. 2.5 Characteristics of micro-fractures in thin sections under microscope. **a** XS301well, volcanic reservoir rhyolitic ignimbrite of the first member of Yingcheng Formation in Xudong area, Songliao Basin, weak dissolution of CO_2-containing acidic water, 907.21 m; **b** XS12 well, the first member of Yingcheng Formation volcanic reservoir in Xudong area of Songliao Basin, 3666.18 m; **c** J2 well, the intergranular pore of fan delta front sedimentary reservoir in a certain area of western sag of Liaohe Basin, depth: 959.82 m, 50×. **d** J2 well is the intergranular pore of fan delta front sedimentary reservoir in a certain area of western sag of Liaohe Basin. The depth is 974.56 m, 25×

(5) Physical simulation

Physical simulation is the most widely used technique for geological genetic analysis of structures. It involves structural evolution simulation, stress field simulation, and fault simulation. The point is that experimental conditions should approach subsurface geologic conditions at different stages of evolution in the geologic history. Li (2001) used tectonic stress field analysis and structure physical simulation experiments for geological genetic analysis of structures. Xue et al. (2017) simulated the structural features and their genetic mechanisms of the Hala'alate Mountains and primarily used physical simulation to establish the tectonic deformation processes at typical sections in different zones of the study area. In China, physical simulation of a fault system can be performed at the China University of Petroleum-Beijing and Northeast Petroleum University. Physical simulation experiments dynamically reproduce the stages and processes of structure development for a better direct understanding of the geologic genetic mechanism of structures. However, it hardly works and is risky at high temperature and high pressure. Additionally, the experimental process is often distinct from the actual geologic history, which is extremely difficult to address.

(6) Well logging interpretation

Well logging interpretation is also a common technique for geological genetic analysis of structures. Zhang et al. (2014) used imaging logging, core, and thin-section data to identify structural fractures in the Yubei area of the Tarim Basin; classified the stages of fractures according to the fracture intersection, cathodoluminescence of interstitial calcites, and inclusions; used the finite-element method to numerically simulate the stress field of paleostructures at different stages of fracture; and established fracture genetic models dominated by faults and folds depending on the stress field and mechanical mechanism at different stages. Chen et al. (2011a, b, c) used imaging logging and core observation and correlation in addition to conventional well logging interpretation to analyze the relation between fracture and lithofacies for fracture characterization. Vertical, high-angle, low-angle, and horizontal fractures have different imaging log responses, especially stage-3 fractures. Microfractures of 1–40 μm wide can be directly imaged in FMI curves. Log data can also be used to calculate the fracture index and fracture width. There are many types of log data that can be easily measured for qualitative and quantitative fracture characterization, but it is difficult to obtain accurate interwell prediction. Additionally, log-based fracture interpretation is liable to be affected by the researcher's practical experience and instrument precision, the latter of which needs to be corrected if necessary.

(7) Seismic interpretation

Seismic interpretation is the most commonly used technique for tectonic origin analysis. Well and seismic data are jointly used for section-based fracture interpretation and horizon tracking followed by the grouping of fault points on the map. The fault system obtained through seismic interpretation is combined with the geologic setting of the target horizon for geological genetic analysis of structures (Fig. 2.2). Li et al.

(2015a, b) investigated Cenozoic structures and their genetic mechanisms in the eastern Liaodong Bay Depression based on 3D seismic data, and they classified the structural evolution in this region into several stages, that is, weak strike slip and strong extension in the Paleocene and Eocene, weak extension and strong strike slip in the Oligocene, and weak compression and weak strike slip in the Neogene. Xu et al. (2016) used seismic data to work on Cretaceous salt structures and their genesis in the North Gabon Subbasin. The seismic interpretation has also been widely applied to the geological genetic analysis of structures in field development because of its advantage in relatively accurate interwell structure prediction, particularly based on three-dimensional (3D) high-precision seismic data. However, seismic survey suffers from high cost and the limitation of natural conditions and high-precision seismic data are not available for all blocks of interest. In the Changqing Oilfield, the Ordos Basin, for example, the structure study mainly relied on well data instead of seismic interpretation. Additionally, seismic interpretation should be performed after well-tie calibration. Structural evolution and in situ stress orientation should be considered comprehensively in structure interpretation, especially fault point grouping. In other words, seismic interpretation is closely related to the interpreter's experience and the practice in the area of interest.

(8) Lab test and statistics

There are many lab test and statistical methods related to the geological genetic analysis of structures, such as scanning electron microscopy (SEM), X-ray diffraction, water composition analysis, fluid inclusion dating, and isotopic dating. SEM can be used to observe the structural features of microfractures. The difference in the water composition between the upthrown wall and downthrown wall of a fault may indicate fault sealing and its geological genesis. The author used SEM data to identify microfractures in glutenite reservoirs in the Lower Karamay Formation at the northwest margin of the Junggar Basin (Fig. 2.6) and obtained direct results of the size, geometry, and the location of microfractures. These results may facilitate enhanced recovery in oil/gas fields.

(9) Geological modeling

Geological modeling is a proven technique commonly used for geological genetic analysis of structures. Wall et al. (2004) focused on the age, origin, and tectonic significance of Mesozoic sediment-filled fissures in the Mendip Hills on the northern margin of the Wessex Basin, SW England, using predictive models and Jurassic sea-level curves. They concluded that the fissures occur in the Lower Triassic to Jurassic between the Carboniferous limestone and the unconformably overlying the oolitic limestone of the Lower Bajocian. The Upper Triassic–Lower Jurassic fissure-fill stratigraphy in the eastern Mendip area was constructed based on cross-cutting relationships, sedimentary facies, biostratigraphy, lithostratigraphy, and strontium-isotope ages of relevant Mesozoic sediments. Deibert et al. (2006) used the predictive model to reveal the sedimentological and tectonic origin of an incised valley along an extensional marginal lacustrine system in the Knoll Basin and Range Province,

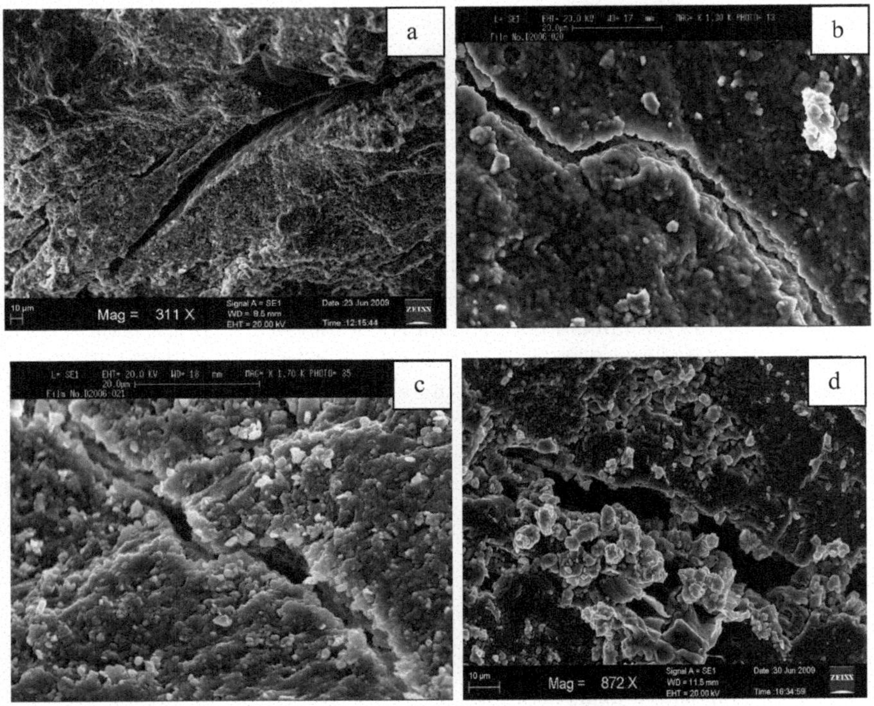

Fig. 2.6 Scanning electron microscope characteristics of micro-fractures in alluvial fan sedimentary reservoir of Kexia Formation in a certain area of northwestern margin of Junggar Basin. **a** J3 well, micro-fracture, brown-gray oil-bearing sandy fine conglomerate, 396.09 m; **b** T10 well, microfracture, brown-grey small conglomerate, 1069.64 m; **c** T10 well, micro-fracture, gray glutenite, 1101.15 m; **d** J3 well, intergranular fracture and oil immersion, gray grained coarse sandstone, 400.80 m

Nevada, the United States. Thus, the origin of the incised valley could be better understood from the perspectives of sedimentation and tectonism. The author predicted interwell structures through geological modeling during geological genetic analysis of structures of the Yulou oil reservoir in the steam-drive test area, the West Sag of the Liaohe Basin (Fig. 2.7). Abundant well data (due to well spacing interval generally of 300 m) and 3D high-precision seismic data mostly available during field development lead to credible detailed structure interpretation using well and seismic data and consequent accurate credible geologic models. The geological model should be built based on enough geological study; otherwise, it may only be a digital game and cannot offer support to production. There are many problems in geological modeling of fractures at present, which may be solved by improving algorithms and techniques.

(10) Dynamic monitoring and production performance analysis

Abundant dynamic monitoring and production performance data during oil/gas field development facilitate geological genetic analysis of structures at this stage. Tracers

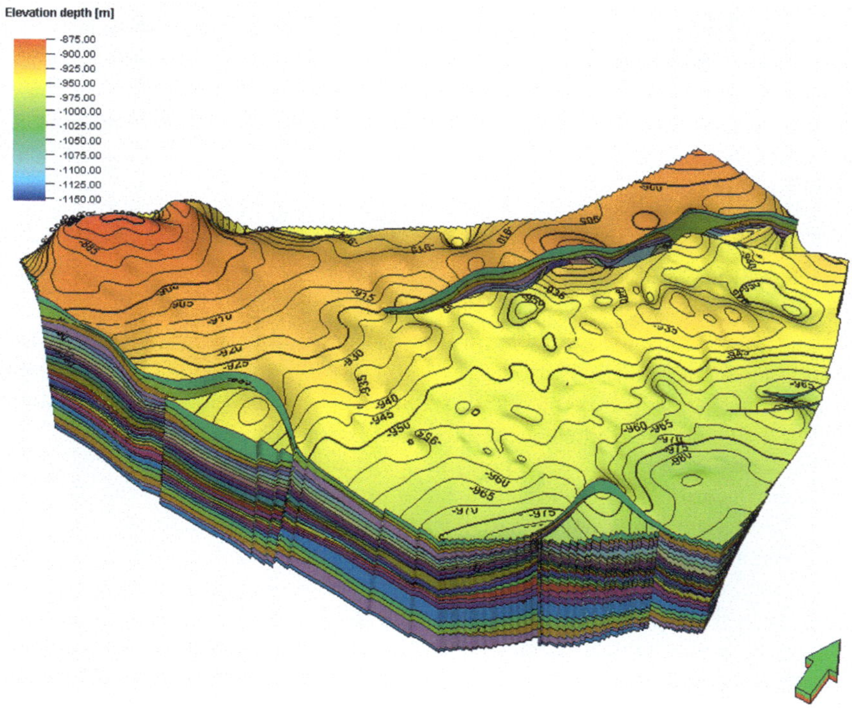

Fig. 2.7 Structural model of Yulou oil layer in a certain area of western sag in Liaohe Basin

are used to monitor the change between the tracer injection well and the well being monitored and thus to estimate interwell fault sealing. Production performance data of injectors and producers can be used to analyze injector-producer correspondence and various geologic factors, for example, the occurrence, size, and density of fractures and the occurrence and sealing the property of faults, which may affect injection-production correlation. Note that production performance depends on many factors instead of only the tectonic factor, which needs to be diagnosed comprehensively using other techniques. Despite its advantages in abundant data, close relation to production, and easy application to production, dynamic monitoring and production performance analysis suffers from large uncertainties, which may be mitigated using various techniques including basic geological analysis.

The geological genetic analysis of structures is a systematic task involving several components. In view of multielement impacts on geological genetic analysis of structures, proper techniques, with distinct advantages and disadvantages, should be optimized in accordance with data foundation and the target to be investigated to complement each other and arrive at good results.

2.1.4 Problems and Prospects of Geological Genetic Analysis of Structures

Geological genetic analysis of structures is the most fundamental work in oil/gas field development, and it has greatly progressed over the decades. This section presents the problems and prospects thereof.

2.1.4.1 Problems

(1) Research findings are biased by insufficient integration with sedimentological and diagenetic studies. Pierre et al. (2007) reported the tectonic, sedimentary, and diapiric processes in the chaotic sediments from the Messinian of the Torino Hill, the Paleogene and Neogene Piedmont Basin, northwestern Italy.

(2) Qualitative investigation based on geological and geophysical techniques overwhelms quantitative investigation. Sàbat et al. (2011) examined the origin of the Western Mediterranean based on the geological structure and evolution of Majorca in eastern Spain and concluded that thrust faults dominate over Majorca. Fault delineation was mainly performed using seismic sections.

(3) Geological genetic analysis often relies on and is restricted to a simple method that has its limitations. Leeuwen et al. (2007) used petrologic and isotopic analysis to approach the origin and tectonic history of the Malino Metamorphic Complex, NW Sulawesi, Indonesia. Morteani et al. (2010) discussed the amethyst geodes at Artigas in Uruguay and the paleohydrology of the Guaraní aquifer from the structure, geochemistry, oxygen, carbon, and strontium-isotope perspectives and fluid inclusion.

(4) Geological modeling, especially geological modeling of fractures, leaves much to be desired, with uncertainty in the accuracy of spatial prediction. Warren et al. (2007) focused on the structural and stratigraphic controls on the origin and tectonic evolution of a subducted continental margin in Oman based on a revised tectonic model, which explains the structural, geochronological, and metamorphic observations.

(5) Existing techniques for geological genetic analysis of structures are inadequate and should be improved along with related disciplines. New techniques and methods are expected.

(6) Macroscopic studies overwhelm microscopic studies in the geological genetic analysis of structures. Fracture characterization is performed mainly using core, thin-section, SEM, and other lab test data.

(7) Geological genetic analysis of structures mainly focuses on clastic rocks and seldom on carbonate, volcanic, and metamorphic rocks. Reservoirs of different lithologic geneses are significantly variable in the geological genesis of structures; thus, they should be investigated with specific techniques and methods.

(8) Geological genetic analysis of structures needs to be expanded to show its significance and value. It is mainly applied to fault interpretation and fracture analysis at present, and its significance to fault-sealing evaluation, structure-reservoir architecture characterization, and remaining oil characterization based on microstructures has not been fully realized.

2.1.4.2 Prospects

(1) Geological genetic analysis of structures is combined with sedimentological and diagenetic studies for a comprehensive diagnosis of tectonic origin. Lavenu et al. (2013) probed into the tectonic versus diagenetic origin of fractures in a naturally fractured carbonate reservoir along the Nerthe anticline in southeastern France.
(2) More quantitative efforts, such as lab tests of in situ stress in petrology and quantitative calculations of fracture index using log data, are taken for geological genesis analysis of structures, while the qualitative investigation is strengthened.
(3) Certain techniques are integrated for a comprehensive geological genetic analysis of structures to achieve optimal results. Wu et al. (2015a, b) used drilling, seismic, geological, gravity, and magnetic data for structure interpretation of basement rifts to establish geometric and kinematic features and a genetic model of the Gaoyou Sag in the Subei Basin.
(4) The geological modeling algorithm is optimized to expand the technique's application in the geological genetic analysis of structures. Neill et al. (2010) structurally reconstructed the origin of the volcanic complexes of La Désirade in the Lesser Antilles with the aid of a prototype model of the Late Jurassic to Cretaceous Pacific Caribbean margin.
(5) While existing techniques are improved, new techniques and methods are figured out for geological genetic analysis of structures. Dufréchou et al. (2013) established tectonic models for the origin of regional transverse structures in the Grenville Province of SW Quebec, Canada, interpreted from regional gravity.
(6) Microstructures and fractures are incorporated to improve the accuracy of geological genetic analysis of structures.
(7) Geological genetic analysis of structures is extended to reservoirs of other lithologies, such as carbonate, volcanic, and metamorphic rocks. Cengiz et al. (2010) discussed the origin of Neogene tectonic rotations in the Galatean volcanic massif, central Anatolia and concluded that two-phase rotations may be the product of Neogene volcanic activity and that the origin, type, geometry, and other properties of fractures were related to the geodynamic history. Klimchouk et al. (2016) investigated the hypogenic origin, geologic controls, and functional organization of a giant cave system in Precambrian carbonates, Brazil, and suggested that the formation of the cave pattern was controlled by a system of subparallel anticlines and troughs with a NNE-SSW dominant orientation and that the cave distribution was spatially controlled by the rift system.

(8) Geological genetic analysis of structures is gradually promoted to other studies on fault sealing, reservoir architecture, microstructures, and remaining oil to offer greater support to oil/gas field development.

2.2 Fault Analysis in Field Development

2.2.1 *General*

A fault is a basic tectonic style occurring widely in the crust and is closely related to the richness of minerals and energy resources (Zhuang et al. 1996). Many scholars have devoted great efforts to its research (Liu et al. 1998a, b; Sun et al. 2000; Stenger et al. 2002; Lv et al. 2002, 2013; Shen et al. 2003; Luo et al. 2007; Braathen et al. 2009; Ameen et al. 2010; Chen et al. 2010a, b; Zhang 2010; Zhang et al. 2017; Feng et al. 2017). Liu et al. (1998a, b) presented the latest progress in fault study from the perspectives of mechanical mechanism, constraints, major fault types, and methodology. Sun et al. (2000) discussed the orderliness of faults and classified them into mantle faults, crust faults, and formation faults. Stenger et al. (2002) investigated fracture reactivation dominated by injection pressure in the Upper Jurassic Arab-D reservoir, the Ghawar Field, Saudi Arabia, and suggested that fracture re-opening may be related to the neotectonic activity with pore pressure recovery in a period of time, which was supported by seismic events recorded in and around the study area. Lv et al. (2002) wrote *Research on Fault Sealing*, the first monograph that elaborates research on fault sealing in China. Shen et al. (2003) dealt with the application of fluid inclusions to fault studies, involving faulting and hydrocarbon migration, hydrocarbon preservation, and fault structure. Luo et al. (2007) elaborated the mechanism and model of fault controlling petroleum accumulation with several examples from the Bohai Bay Basin in East China to the Tarim Basin in West China. Braathen et al. (2009) proposed a novel concept of fault facies for fault description adapted to 3D geological modeling of sandstone reservoirs in Sinai, Egypt. Fault facies are related to field data on the dimensions, geometry, internal structure, petrophysical properties, and spatial distribution of fault elements. Ameen et al. (2010) studied the static and dynamic data from the Permian–Triassic Khuff reservoir in 19 traps, onshore Saudi Arabia, to identify fractures and in situ stress and concluded that in situ stress was controlled by the Zagros plate tectonics and that the presence of fractures has a great impact on productivity. Chen et al. (2010a, b) classified the pathway system of the Paleogene Lingshui Formation in the Qiongdongnan Basin into the following categories: reservoir, unconformity surface, and fault. Zhang (2010) used geochemical, physical, and numerical simulation techniques to elaborate fault-controlled oil reservoirs and hydrocarbon accumulation in the Junggar Basin. Lü et al. (2013) detailed the effects of faults on hydrocarbon migration and accumulation. Zhang et al. (2017) analyzed the different sealing conditions required by mudstone caprock at different stages of fault evolution with a case study on the first member of the Qingshankou

Formation in the Sanzhao Sag and the second member of the Dongying Formation in the No. 5 structure of the Nanpu Sag. Feng et al. (2017) used measured stress data and other stress data derived from the focal mechanism solution, borehole collapse, stress relief, and inverted fault slip vector version to analyze the present-day tectonic stress field at the shallow crust near the Tanlu fault zone. More accurate fault analysis is required in field development than in petroleum exploration. Such investigations may be realized through geological genetic analysis in addition to improved geologic and geophysical techniques. Previous fault studies dealt with the genetic mechanism of the fault system, association and classification of the fault system, fault sealing, and quantitative characterization of fractures of different genetic types using geochemical methods, physical and mathematical simulations, outcrop and core observation and description, thin-section observation, seismic and logging data interpretation, and production performance monitoring and analysis. This book demonstrates the relation between fault analysis and field development and confines the targets to crust and formation faults, including faults and fractures (Wu et al. 2005; Yang et al. 2009). A fault system occurs extensively in petroliferous basins in China, from the Bohai Bay Basin in the east to the Ordos Basin in the middle and to the Sichuan and Junggar Basins in the west. With respect to oil/gas field development, taking CNPC as an example, there are many problems to be solved in fine fault interpretation and fault sealing identification due to the limitation of data quality and the level of research, and fault analysis has not been taken as seriously as reservoir characterization. Thus, based on the fault analysis for field development in previous studies and in practices with respect to volcanic gas reservoir development planning, glutenite reservoir architecture analysis, and thermally recovered heavy oil reservoirs, this book expounds fault analysis in field development, as well as the problems and focuses of research in detail.

2.2.2 *Foundations of Fault Analysis in Field Development*

2.2.2.1 Data

Fault analysis is a systematic work in field development. It involves many components, including well-seismic fault interpretation and fracture characterization, which should be based on the proper understanding of basin structural evolution, in situ stress field, sedimentary facies, and diagenetic evolution that are related to the geological genesis of the fault system. Relevant data mainly include basic geological, outcrop, well drilling, seismic, lab test, dynamic monitoring, and production performance data. Figure 2.8 shows the outcrop responses of fault systems in various basins, which reflect clastic, carbonate, and volcanic oil reservoirs. Fault analysis is usually target-oriented and cross-validated using different types of data. Indeed, more data does not necessarily lead to good results. The data to be used should be properly optimized to avoid excessive workload and some mistakes. With respect to data application, outcrop data are the most direct for fault recognition. Through

Fig. 2.8 Characteristics of fault systems in field outcrops. **a** High-angle fractures in the Triassic Yanchang Formation, Ordos Basin (Yanhe section); **b** field outcrop characteristics of fault system in Ji County of Tianjin City, caves in limestone in the Tieling Formation; **c** characteristics of Tiankeng Fault System in Wulong County of Chongqing City; **d** faults in the purple-grey Mesozoic breccia lava tuff in Yanqing District of Beijing City

outcrop observation and description, researchers can have an insight into the pattern of the fault system and thus offer support to field development planning and stimulation treatments. Well and seismic data are most widely used in fault analysis to quantitatively delineate the size, location, and properties of the fault system. Lab test data are also useful in fault analysis. For example, oilfield water analysis data can be used to diagnose fault-sealing capacity. Dynamic monitoring and production performance data have also been useful in fault analysis. On the one hand, they are dynamic evidence of the fault system as a complement to conventional static studies. Alternatively, fault analysis based on a combination of dynamic and static data is more practical and applicable.

2.2.2.2 Techniques

Diverse techniques are inevitable for fault analysis in field development, where different types of data are used depending on the target of interest, as described above.

The techniques include basic geological analysis, outcrop observation and description, core observation and description, thin-section observation, physical simulation, well logging interpretation, seismic interpretation, lab tests and statistics, geological modeling, and dynamic monitoring and production performance analysis. Each technique has its unique features and capacity. For example, outcrop observation and description underline the geologic correlation between development areas and outcrops. As a simple and feasible practice for this purpose, an outcrop at the basin margin with the same stratigraphic horizons as the development area is found and geologically correlated with downhole conditions in the context of basically consistent features, such as in situ stress; then, the derived outcrop tectonic geology is properly extended to the development area for analyzing the structural features in the target horizon. Well logging interpretation and seismic interpretation are the most commonly used techniques in fault analysis. Well logging interpretation yields the fracture porosity, fracture permeability, fracture index, and fracture width for quantitative fracture characterization. The seismic interpretation provides fine recognition of faults (the third-order fault or the fourth- or lower-order faults in local areas where the target horizon is shallow and seismic data quality is good) after calibrating with data from coring wells and being constrained by fault points derived from development wells, thereby facilitating remaining oil characterization and enhanced oil recovery. Dynamic monitoring and production performance analysis is generally combined with the above static techniques. Overall, fault analysis techniques are becoming more comprehensive, quantitative, and multitudinous.

2.2.3 *Fault Analysis in Volcanic Gas Reservoir Development Planning*

2.2.3.1 Well-Seismic Fault Interpretation

Fault analysis is crucial to volcanic gas reservoir development planning. As for volcanic gas reservoirs in the Ying-1 Member in the Xudong area, the Songliao Basin, fissure–central eruption is in the majority, followed by central eruption (Chen et al. 2012a, b). The volcanic crater and ancient landform spatially control the occurrence of various volcanic facies, which are closely related to reservoir properties. This means that crater location is closely related to reservoir properties. Craters mostly occur along large faults cutting to the basement; in other words, locating large faults means locating high-graded volcanic reservoirs. Thus, accurate delineation of large faults is geologically important for high-graded volcanic reservoir prediction and development planning. Fractures associated with large faults may connect effective storage space (e.g., vesicles) in rhyolites to improve petrophysical properties and form high-graded reservoirs. Therefore, fine fault interpretation is a major focus in high-efficiency volcanic reservoir development. The above conclusions have been confirmed by actual gas field development. High-graded reservoirs

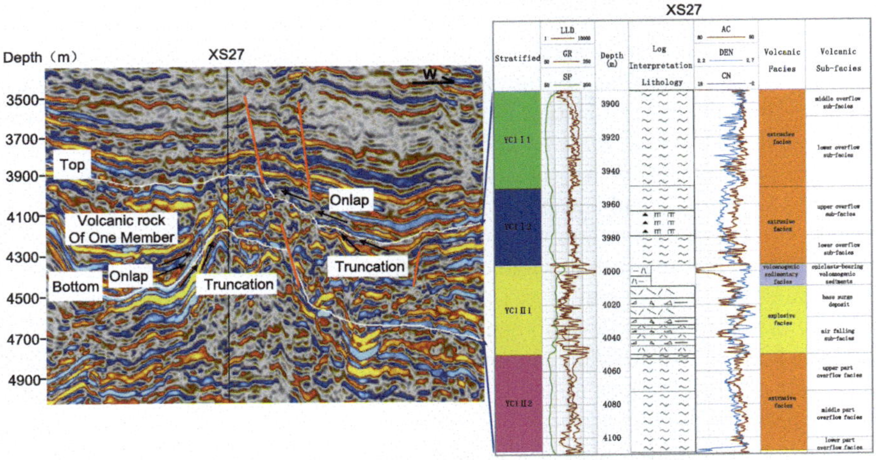

Fig. 2.9 Sectional development characteristics of faults from interpretation by combining well data with seismic data in the Member One of the Yingcheng Formation in the Xudong Area, Songliao Basin

with large gas deliverability are usually drilled at the structural highs close to craters. The author finished well-seismic fault interpretation for volcanic reservoirs in the Ying-1 Member in the Xudong area, the Songliao Basin (Fig. 2.9). For the purpose of fault interpretation, the researchers need to fully understand the structural history of the basin, durations and scales of tectonic activities, and in situ stress orientation. Moreover, well-seismic consistency should be guaranteed in addition to accurate well-tie calibration. Generally, it is required to correctly group over 80% of fault points identified at well sites in well-seismic fault interpretation. Based on data of close-spaced wells and 3D seismic data, fine fault interpretation and description can be accomplished for structure modeling to characterize the spatial distribution of the fault system in the target horizon. According to Li (2007), who classified faults into five orders, the third- and higher-order faults can be accurately delineated using seismic and well data, with a genetic analysis of the fault system, whereas the low-order (fourth- and fifth-order) faults cannot be accurately interpreted merely depending on well and seismic data. In the latter case, tracer monitoring and production performance data should also be integrated to alleviate the uncertainty of seismic interpretation.

2.2.3.2 Fine Fracture Characterization

As mentioned above, for volcanic reservoirs in the Ying-1 Member in the Xudong area, the Songliao Basin, fractures mainly turn up close to major faults and connect different types of pore space, increasing improved reservoir properties. Thus, final fracture characterization is also an important component of volcanic gas reservoir development planning. It is mainly realized using geology, core, thin section, logging, and seismic data based on the classification of fractures by geological genesis. Figure 2.10 shows fractures observed on thin sections. Some fractures extend long in the reservoirs and effectively connect pore space (Fig. 2.10a, b, c), while some are filled with bitumen (Fig. 2.10d) or silica, leading to deteriorated reservoir properties. Thin-section observation is an intuitive and semiquantitative means of fracture characterization, but it provides a limited amount of data in a small scope. Ant tracking, a module in Petrel, was used for more detailed seismic-based lateral fracture characterization than coherence analysis. As indicated by ant tracking, fractures mainly extend roughly in an N–S direction along the Xuzhong and Xudong faults in the Xudong area. Few large faults concentrate in the northern and central parts of the study area, and many small faults scatter in the southern part (Chen et al. 2016a, b). Microscopically, there are numerous E–W or nearly E–W fractures in addition to the S–N fractures near the fault zones. These fractures spatially constitute a 3D network in the Xudong area and have played an important role in reservoir properties (Chen et al. 2016a, b). These fractures have significantly improved the physical properties of low-permeability volcanic reservoirs and thus become an important control on reservoir properties. Those fractured structural highs should be the major targets for well deployment in development planning. Fracture characterization-using ant tracking is nonunique, and other techniques including coherence analysis and fracture-logging interpretation at a dense well pattern should be combined to obtain accurate results.

2.2.4 Fault Analysis in Glutenite Reservoir Architecture Analysis

Reservoir architecture involves the geometry, dimension, orientation, and stacking of different reservoir units (Wu 2010). For a glutenite reservoir at a high water-cut stage, enhanced oil recovery (EOR) requires a better understanding of reservoir properties, which cannot be achieved through conventional sedimentary microfacies analysis. Reservoir architecture analysis has become an alternative solution. Generally, architecture analysis focuses on sedimentary architecture, that is, further classification of reservoir sedimentation units based on sedimentary microfacies to the level of single sand body to investigate the spatial occurrence of sand bodies of different genetic types. However, actually, there are other types of architectures in addition to the sedimentary architecture. The author divided the glutenite reservoir of the Lower Karamay Formation in an area at the northwestern margin of the Junggar Basin into three types of architectures—sedimentary architecture, diagenetic architecture, and structural architecture, the last of which was caused by early faults filled with silica or bitumen to form barriers (Fig. 2.11). As for multiphase faults, early fracture architectures are commonly filled with silica to form barriers against fluid flow, while late fractures are usually open to function as flowing channels that may be responsible for water breakthrough during production. Thus, fault analysis is important for EOR

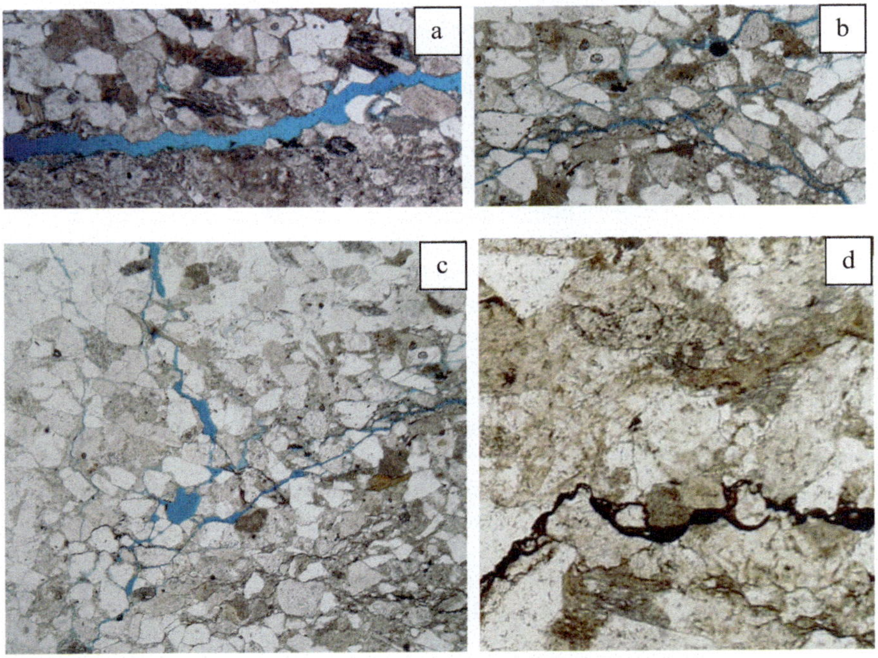

Fig. 2.10 Fracture characteristics of microscopic thin sections in the Member One of the Yingcheng Formation in the Xudong Area, Songliao Basin. **a** Fracture, Well XS23, Depth: 3722.03 m, ×10, (+); **b** fracture, Well XS21, Depth: 3527.88 m, ×10, (+); **c** fracture, Well XS21, Depth: 3532.48 m, ×10, (+); **d** Well XS24, suture in tuffaceous sandstone, Depth: 3688.05 m, ×40, (+)

from glutenite reservoirs at the middle and late stages of development (with high water cut). Figure 2.11a shows high-angle fractures filled with silica, which may form a fracture architecture that functions as barriers against fluid flow. Figure 2.11b shows three high-angle fractures, of which the first two from the left are late unfilled fractures that cannot be taken as barriers and the rightmost one is filled with silica and may function as a barrier against fluid flow. The fracture architecture can be clearly observed on cores, but its dimension can hardly be estimated owing to the small size of cores. The genetic type, dimension, and the effect on oil and gas production of the fracture architecture may be further analyzed using well-seismic fault interpretation and production performance data.

Here, a well cluster consisting of five wells (one oil producer T6070 and four water injectors T6060, T6080, T6059, and T6082) are selected to demonstrate the effect of the architecture on reservoir connectivity (Fig. 2.12). As per production performance data, the producer is not efficiently swept after water injection at the four injectors, which means that reservoirs are poorly connected between the injectors and the producer. Geological analysis shows that the pay zones S732 and S733 lithologically consist of laminar-flow gravels at five wells, which extend in a large area. Therefore,

Fig. 2.11 Cores photographic characteristics of fault architecture in the glutenite reservoir of the Lower Karamay Formation in a block at the northwestern margin of the Junggar Basin. **a** Well J581, high-angle fracture, light brown medium gravel glutenite-light gray green mudstone, 393.27–393.48 m. **b** Well J581, high-angle fracture, light gray green pebbled medium-fine sandstone, 406.51–407.5 m

it is difficult to account for poor reservoir connectivity. In accordance with single-well and interwell architectures, there are four orders of sedimentary architecture boundaries in laminar-flow gravels between T6060/T6081/T6059 and T6070, and there is a fracture architecture boundary between T6082 and T6070. This indicates that reservoir connectivity in this well field is dominated by architectures, and it also demonstrates the significance of fault analysis for EOR (Chen et al. 2015a, b, c).

2.2.5 Fault Analysis for Thermally Recovered Heavy Oil Reservoirs

This section discusses fault analysis for thermally recovered heavy oil reservoirs from the perspectives of fault interpretation and modeling, as well as fault sealing.

2.2.5.1 Fault Interpretation and Modeling

Fault interpretation and geological modeling are an important component in the development of thermally recovered heavy oil reservoirs. Through fine fault interpretation, the structural model of the target horizon is built for subsequent reservoir numerical simulation and development planning. In the author's geological study of the Yulou oil reservoir in the West Sag, the Liaohe Basin, the 3D seismic data

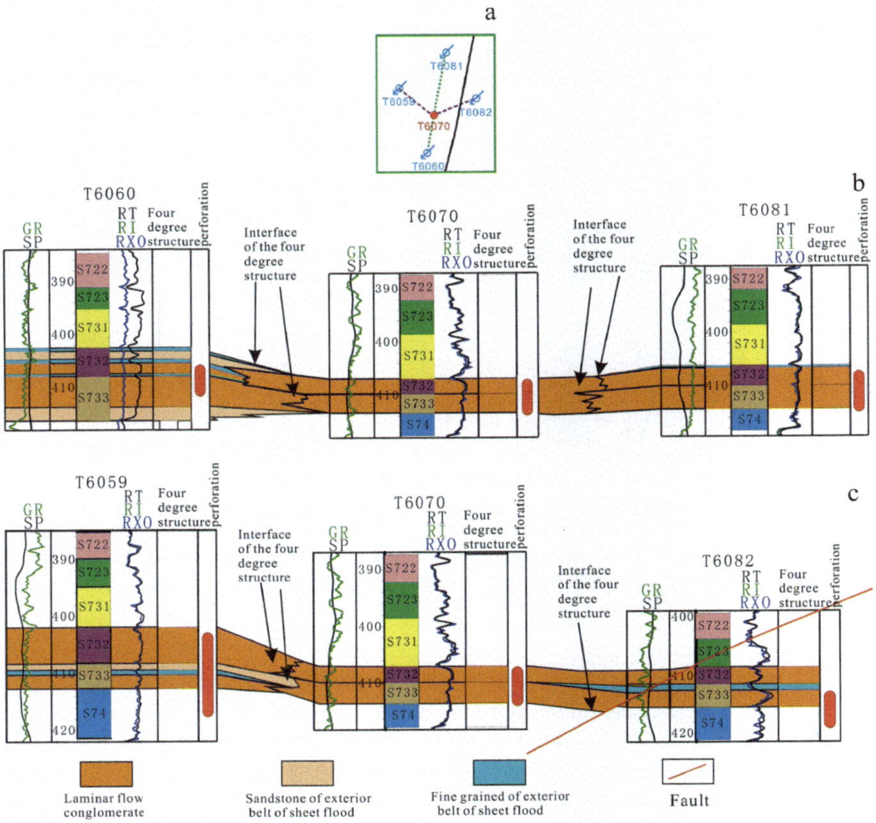

Fig. 2.12 Influence of architecture characteristics on reservoir properties and oilfield development (Chen et al. 2015a, b, c). **a** Horizontal well location map; **b** North–south section **c** East–west section

were calibrated using the well data with core samples for time-depth conversion of seismic data. Then, coupled with the structural evolution and in situ stress field of the basin, faults of different orders were spatially delineated through section interpretation and fault point grouping on the map. Finally, the fault interpretation results were input into Petrel to build the structural model of the target horizon for 3D spatial fracture characterization and subsequent sedimentary microfacies modeling and facies-controlled reservoir attribute modeling. Figure 2.13 clearly illustrates the spatial distribution of faults in the target horizon, which mainly extend in the E-W and NE-SW directions. The most important characteristic and advantage of the geological model is a 3D quantitative characterization and interwell prediction of fractures. Fault modeling is based on fine fault interpretation using well and seismic data, and the agreement between log-interpreted fault points and seismic-interpreted fault

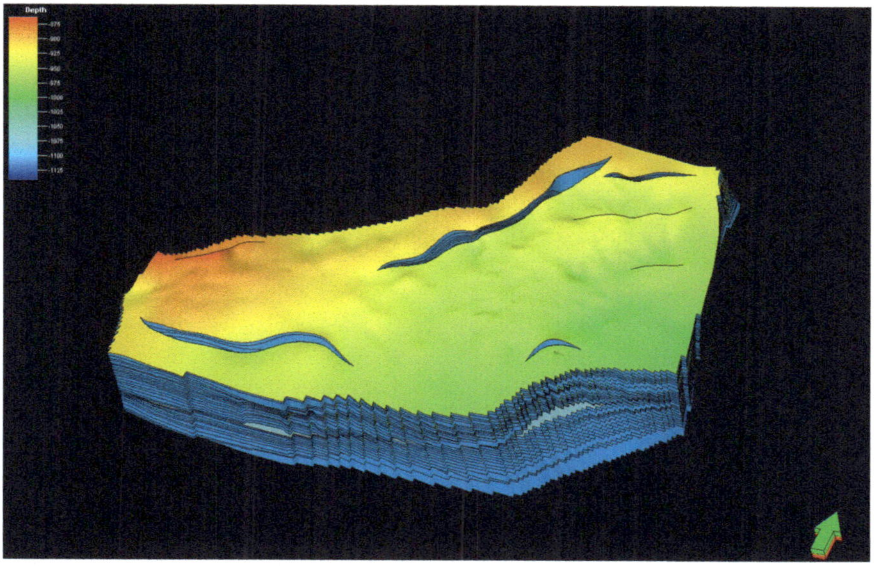

Fig. 2.13 Characteristics of structural geological model of Yulou oil layer in a certain area of western sag of Liaohe Basin

points is key evidence to validate the fault interpretation. In fine reservoir description, it is required to correctly group over 80% of log-interpreted fault points during seismic structure modeling (Chen et al. 2019b).

2.2.5.2 Fault Sealing

The significance of the fault sealing to oil/gas field development is attracting more and more attention from researchers, particularly for the reservoirs in most faulted lake basins in East China, where the fault sealing directly affects reservoir connectivity and injection-production correlation. Wang (1999a, b, c) presented that various faults in oil and gas fields, from extensional normal faults in East China to compressional reverse faults in West China and from large faults of hundreds of meters in throw to small faults smaller than 10 m in throw, have been demonstrated by a waterflood of continental oil reservoirs to function as barriers. However, this is not completely true. Chen et al. (2015a, b, c) used detailed water analysis data to account for lateral and vertical fault-sealing properties in the Yulou oil reservoir on the west slope of the West Sag of the Liaohe Basin. Fault sealing is commonly investigated by lithologic distribution at both sides of the fault, the configuration of the fault occurrence and formation occurrence, correlation between fault orientation and fault-sealing, configuration of fault activity and hydrocarbon migration, fault mechanical properties, drainage pressure along the fault plane and both sides of the fault, log responses of fault points, drilling shows, fluid properties and distribution at both sides of the fault, and production performance analysis (Zhang et al. 2011). Each technique has its data foundation, operational principles, emphasis, advantages, and disadvantages, which should be considered and selected properly depending on data availability and research target. Typically, fault-sealing analysis based on fluid properties at both sides of the fault is an important technique used in field development. In this book, water analysis data at both sides of the fault are used to check fault sealing. There are mainly four faults—F1, F2, F3, and F4 in the study area (Fig. 2.14), of which F1 and F2 are third-order sag-controlling faults and F3 and F4 are fourth-order faults. The water analysis data of 10 wells in 5 groups were compared (Table 2.2) to estimate fault sealing (Chen et al. 2015a, b, c). Large discrepancies in water analysis data at both sides of the fault indicate a good sealing property, while similar results indicate a poor sealing property. For F1, Wells A1 and A2 at both sides show remarkably different water analysis results, especially the Mg^{2+}, Ca^{2+}, SO_4^{2-}, and CO_3^{2-} contents, indicating a sealed fault. For F3, two groups, C1/C2 and E1/E2, show equal or similar $K^+ + Na^+$, Ca^{2+}, CO_3^{2-}, and HCO_3^- contents, and total hardness, indicating connected fluids at both sides and unsealed fault. For F4, two groups, B1/B2 and D1/D2, show great differences in the Mg^{2+}, SO_4^{2-}, and CO_3^{2-} contents, total salinity, and total alkalinity, indicating disconnected fluids at both sides and sealed fault. Among the four faults in the study area, F1 and F4 are sealed faults, and F3 is unsealed (Chen et al. 2015a, b, c). Equal or similar water analysis time is a prerequisite for comparative studies to avoid the misjudgment of fault sealing caused by water flooding. In accordance with the above research results, the influence of fault sealing should be considered in the design of an injector-producer pattern when steam huff-and-pull is converted to steam flooding to guarantee a proper injection-production relationship under the presence of fractures and avoid injected steam to channel along opened fractures (Chen et al. 2015a, b, c).

2.2.6 Difficulties and Hotspots in Fault Analysis in Field Development

Fault analysis may be performed to develop various lithologic oil reservoirs, including sandstone, glutenite, volcanic rock, and carbonate rock, and covers the genesis of the fault system, spatial distribution, scale and pattern of faults and fractures, fault sealing, fracture influence on reservoir properties, multi-information fracture characterization, and 3D geological modeling of fractures. At some level, fault analysis directly affects development planning and various stimulations. In the Daqing Oilfield, for example, development wells used to be deployed far away from faults at the early stage of development; after the study of fault control on remaining oil, many development wells have been successfully deployed close to faults for remaining oil recovery at the middle and late stages of development. Except for the

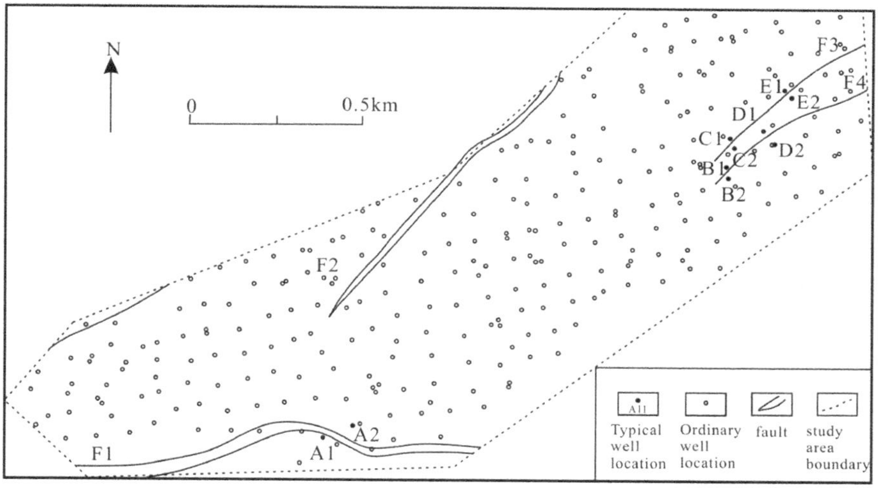

Fig. 2.14 Structural division of Western Depression and location of the study area (Li et al. 2010; Chen et al. 2015a, b, c)

above described, other aspects of fault analysis are bound up with field development. For example, Wang et al. (2015) presented that, in the Changqing Oilfield, natural fractures improve the petrophysical properties of low-permeability oil reservoirs and dynamic fractures may induce a breakthrough of injected water. They further indicated that dynamic fractures reflect the pinnacle water injection profile at some intervals (Fig. 2.15a) and are evidently oriented according to tracer monitoring (Fig. 2.15b). They finally concluded that (1) the dynamic fractures change the seepage characteristics of water displacing oil in the ultra-low-permeability reservoir and greatly aggravate the reservoir heterogeneity, which lead to the reduction of the profile producing degree and the distribution of remaining oil on both sides of the fractures in continuous or discontinuous belts; and (2) the generation and propagation of dynamic fractures are closely related to water injection pressure, injection-production ratio, and various stimulations (Wang et al. 2015). A key problem in deep oil recovery in the Tarim Oilfield is how to effectively integrate the data of widely spaced wells and seismic data to achieve accurate fracture interpretation and delineation. An issue to be addressed in the Daqing Oilfield is how to integrate well, seismic, dynamic monitoring, and abundant production performance data for detailed interpretation of fourth- and lower-order faults. Such problems/issues are important for field development and hot spots in current studies. Researchers engaging in oil/gas field development should make full use of all information and various techniques to achieve fine fracture characterization for high-level oil/gas field development.

Table 2.2 Analytical results of formation water in typical wells of in the Yulou Oil Bearing Sets in a pilot of the Western Depression in the Liaohe Basin (Chen et al. 2015a, b, c)

Well name	Sampling date	Assay date	Na + Ka	Mg	Ca	CL	SO_4^-	CO^{2-}	HCO_3^-	Total mineralization mg/L	Total hardness	Total alkalinity	Watermanshi	PH
A1	2005.06.20	2005.06.21	519.8	4.86	38.1	266	19.21	90	854.28	1792.44	115.1	850.8	$NaHCO_3$	7
A2	2005.06.16	2005.06.17	437	7.3	20	212.8	4.8	0	884.79	1566.69	80.1	725.7	NaH CO_3	6
B1	2010.03.24	2010.03.25	740.6	3.65	10	248.2	24.02	0	1556.01	2582.52	40	1276.2	$NaHCO_3$	6
B2	2010.09.27	2010.09.28	579.6	6.08	12	230.5	14.41	60	1067.85	1970.45	55.1	975.9	$NaHCO_3$	7
C1	2010.10.11	2010.10.12	545.1	9.73	18	230.5	67.24	0	1067.85	1938.45	85.1	875.8	$NaHCO_3$	6
C2	2010.07.14	2010.07.14	503.7	7.3	18	159.6	91.26	0	1037.34	1817.21	75.1	850.8	$NaHCO_3$	7
D1	2001.08.03	2001.08.04	363.4	3.65	6.01	141.8	19.21	150	427.14	1111.25	30	600.6	$NaHCO_3$	8
D2	2001.08.03	2001.08.04	542.8	3.65	12	106.4	24.02	240	793.26	1722.13	45.1	1051	$NaHCO_3$	8
E1	2003.11.02	2003.11.03	446.2	7.3	16	177.3	14.41	90	762.75	1513.99	70.1	775.7	$NaHCO_3$	6
E2	2003.11.12	2003.11.13	407.1	4.86	20	177.3	4.8	90	671.22	1375.32	70.1	700.6	$NaHCO_3$	7

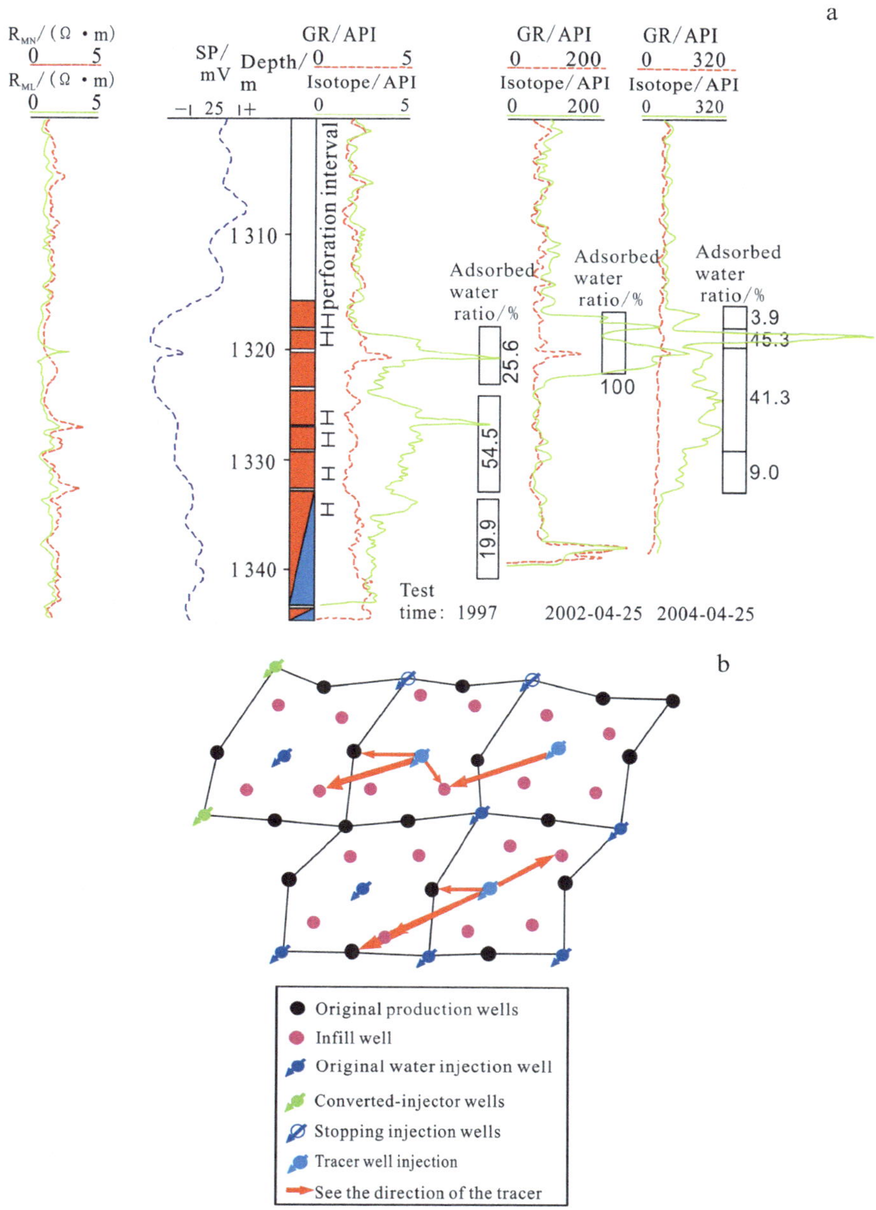

Fig. 2.15 Performance characteristics of dynamic fractures (Wang et al. 2015). **a** Characteristics of water absorption profile evolution; **b** tracer monitoring results

2.3 Multi-information Geological Genetic Analysis and Characterization of Fractures in Volcanic Reservoirs

Fractures are the surfaces of discontinuity caused by rock rupture under force. They may function as storage space and flowing channels in hydrocarbon reservoirs, particularly in fractured reservoirs. Many large and oversized oil and gas fields around the world have been discovered in fractured reservoirs. Systematic studies of fracture type, property, feature, and distribution are significant to the exploration and development of fractured reservoirs, such as volcanic reservoirs (Wu et al. 2005).

Previous efforts into fractures in volcanic reservoirs focused on the following:

(1) Comprehensive fracture identification and prediction using resistivity, sonic, radioactivity, dipmeter, and FMI logs (Deng et al. 2002; Wang et al. 2007a, b, c; Liu et al. 2008).
(2) Fracture characterization mathematically using a hybrid probabilistic method and finite deformation method (Dai et al. 2003; Tang et al. 2009).
(3) Volcanic fissure characterization using the flexural strength method, that is, the rock-rock matrix P-velocity ratio, based on seismic data and attributes (Nie et al. 2007; Zhang et al. 2007a, b, c).
(4) Comprehensive fracture prediction using numerical simulations combined with geology, development, and production performance data (Yang et al. 2010).
(5) Fracture characterization for field development. Wang et al. (1999a, b, c) used a portfolio of fracture characterization techniques, including field survey of similar outcrops and core observation, numerical simulations of the tectonic stress field for fracture prediction, lab tests and analysis, comprehensive analysis of sedimentary microfacies, key well study and multiwell evaluation, dynamic reservoir analysis, drilling engineering analysis, multivariate statistics, neural network simulation and prediction, fractal prediction, geostatistical prediction, fluid flow geologic analysis, and fractured oil reservoir modeling, for fracture description in the Huoshaoshan Oilfield, the Junggar Basin.
(6) Fracture study using isothermal remanent magnetization and thermal demagnetization analysis (Zhang et al. 2007a, b, c).
(7) Genetic mechanism of fractures by correlating fractures with reservoirs for hydrocarbon exploration and development (Li et al. 2004a, b; Liu et al. 2003a, b). d'Huteau et al. (2001) investigated a fissured tuff system of the Upper Cretaceous Castillo Formation in the San Jorge Basin, Argentina, and attributed the low post-frac production to hydraulic fractures parallel to the fissure plane or poor connection between induced fractures and the wellbore caused by premature screen out owing to the existence of multiple fractures.
(8) Fracture geological modeling and prediction. Wu et al. (2015a, b) probed into fracture modeling and prediction through a case study of fractured sandstones with extremely high pressure and low porosity in the Dina two gas fields and

demonstrated that the equivalent parameter model optimized using dynamic parameters yields a fracture attribute modeling with very high accuracy.

(9) New techniques of fracture analysis in laboratory. Gou et al. (2019) employed micron-CT for shale microfracture analysis.

(10) Fracture characterization for unconventional reservoirs. Shang et al. (2021) systematically discussed the application and prospects of fracture characterization and modeling techniques for shale gas reservoirs.

This section introduces the fracture characterization and genetic analysis of the volcanic gas reservoir in the first member of the Cretaceous Yingcheng Formation, Xudong area, the Songliao Basin through the core fracture observation, thin-section fracture observation, log-based fracture interpretation, and seismic-based fracture prediction to offer geological support for effective development of volcanic gas reservoirs.

2.3.1 Geological Setting

The Xushen gas field, 45 km long in the N–S direction and 10 km wide in the E–W direction, lies in Daqing and Anda, Heilongjiang Province. It is structurally located in the Xujiaweizi Fault Depression (formed the Late Jurassic to the early stage of the Early Cretaceous), northern Songliao Basin, and stratigraphically divided into the Huoshiling Formation, Shahezi Formation, Yingcheng Formation, Denglouku Formation, and the first and second members of the Quantou Formation, from the bottom up. Because of frequent volcanic effusive activities, the Yingcheng Formation was settled with mass volcanic products. Volcanic reservoirs, which are mainly acidic effusive rocks, arise from the first and third members of the Lower Cretaceous Yingcheng Formation. There are 69 wells in the gas field, 38 wells of which have yielded commercial gas flow. The gas field is confirmed with gas reserves at the level of 100 billion m^3, 89.8% endowed in volcanic reservoirs, making them the major target of gas development in the Daqing Oilfield (Wu et al. 2002a, b; Wang et al. 2009). The Xudong area, the study area, lies in the middle of the Xujiaweizi Fault Depression, and Well XS1, the discovery well of the Xushen gas field, was drilled in this area. The Xudong area has become one of the most important gas prospects in the Xushen gas field. The target horizon is the volcanic rocks in the first member of the Cretaceous Yingcheng Formation. Fracture analysis of the target horizon is of practical significance to efficient volcanic gas reservoir development in the Xushen gas field and of referential significance to volcanic gas reservoir development in the Songliao Basin and other basins in China (Fig. 2.16).

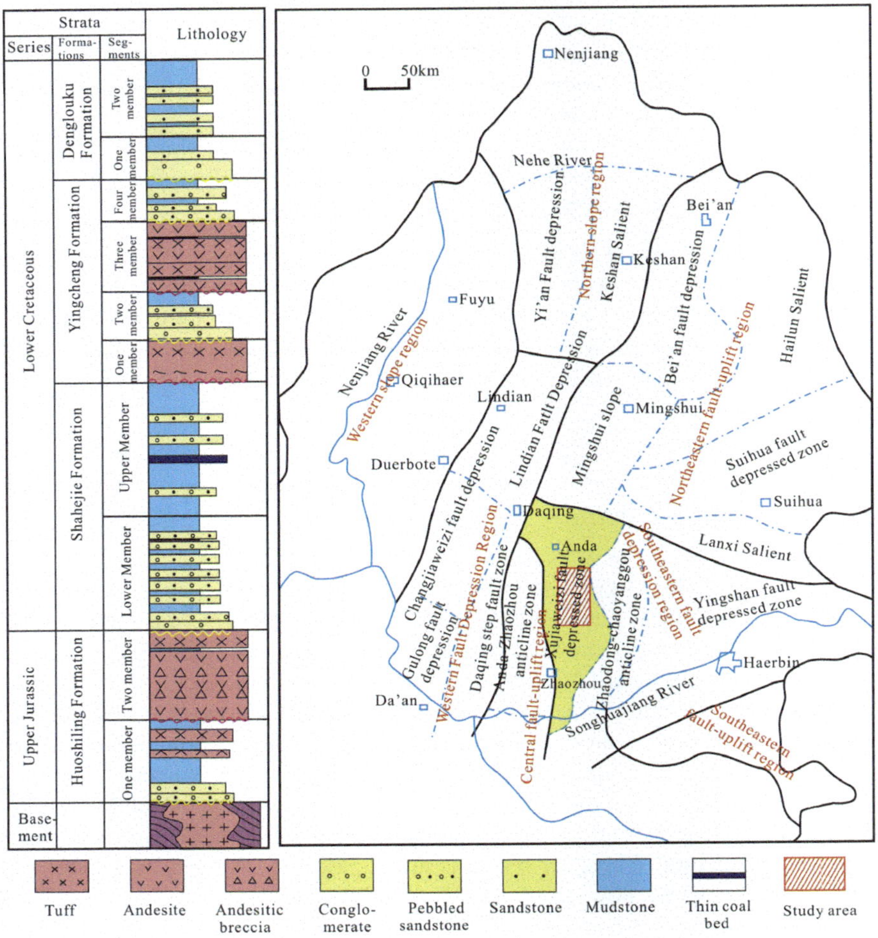

Fig. 2.16 Sketch map of stratum and structure divided of Xushen Gas Field in Songliao Basin (Chen et al. 2016)

2.3.2 Methodology

Based on the identification, classification, and genetic analysis of fractures using core, thin-section, FMI logging, and conventional log data (Chen et al. 2011a, b, c), seismic structure interpretation and coherence analysis were performed to delineation planar fractures in the volcanic gas reservoir. Then, the faulting history and its influence on fractures were determined depending on the structural history and burial-diagenetic history of the basin to understand fracture evolution and spatial distribution of porosity and permeability. Finally, the fractures were classified by development phases, and their influence on/contribution to reservoir performance and their geological characteristics in reservoir development in different areas were analyzed, which made fracture evaluation more specific and referential to reservoir development. Additionally, fractures were analyzed by ant tracking in Petrel based on well and seismic data, which is more detailed than coherence analysis.

2.3.3 Fracture Classification

In view of genesis, fractures in the volcanic gas reservoir in the Ying-1 Member in the Xudong area, the Songliao Basin are classified into structural fractures, condensing shrinkage fractures, burst fractures, dissolution fractures, stylolitic fractures, and weathered fractures (Table 2.3). Structural fractures are induced by or associated with local tectonic actions and are mainly related to faults and folds. Their orientation, distribution, and generation are all related to local structures. Most structural fractures are directional and occur in groups with long extension and large-cutting depth. Despite their small storage space, structural fractures can connect various pores and thus function as flowing channels in volcanic reservoirs to improve reservoir properties. Structural fractures are generally 0.01–0.1 mm wide, and some wide ones are partially or completely filled with calcite or quartz. Structural fractures in the target horizon in the study area are mainly fault-related. Condensing shrinkage fractures are a type of diagenetic fractures caused by volume contraction of effused magma that are cooled and solidified at the earth's surface and usually occur in lava and pyroclastic lava (e.g., rhyolites and brecciated lava), as well as pyroclastic rocks. Burst fractures are the product of gas–liquid explosion of detrital materials in magma during a volcanic eruption and may occur in various volcanic rocks, especially in tuff lava and crystal tuff. Dissolution fractures are generated by corrosion in the original cracks. Net-shaped fractures formed by the dissolution of the matrix in rhyolitic volcanic breccia and secondary fractures created by corrosion between particles of volcanic breccia are wide and effective but small in number—only 10% of structural fractures were corroded, according to statistics. Stylolitic fractures are saw-toothed, and those in the target horizon incise lava phenocrysts and matrix or volcanic debris in volcaniclastic rocks. Most stylolitic fractures are partially or completely filled with iron and mud. Stylolitic fractures are occasionally observed in tuffs and volcanic breccia but have not been examined in other types of volcanic rocks. Weathered fractures are the product of various surface or near-surface mechanical and chemical weathering processes (freezing–thawing cycle, small-scale rock disintegration, mineral alteration, and diagenesis) and mass wasting. Condensing shrinkage fractures can be observed in mudstones (Wu et al. 2005), while burst fractures only arise in volcanic rocks; remaining fracture types are also observed in clastic and carbonate rocks, and they are common in fracture classification by genesis. The classification in this book fully considers the features of volcanic rocks and is more comprehensive than routine schemes.

Table 2.3 Classification of fractures in the volcanic reservoir in the Ying-1 Member in the Xudong area, the Songliao Basin

Feature class	Dimension	Genesis	Fracture density	Host rock	Data foundation
Structural fractures	Large	Tectonism	Large	Various rock types	Core photos and seismic data
Condensing shrinkage fractures	Small	Volcanic eruption and diagenesis	Locally large	Lava and pyroclastic lava	Thin sections
Burst fractures	Small	Volcanic eruption	Locally large	Tuff lava and crystal tuff	Thin sections
Dissolution fractures	Medium to small	Diagenesis	Large	Various rocks	Thin sections
Stylolitic fractures	Small	Diagenesis	Small	Tuff and volcanic breccia	Thin sections
Weathered fractures	Varied	Diagenesis	Small	Volcanic breccia	Core photos

2.3.4 *Multi-information Characterization of Fractures in the Volcanic Reservoir*

2.3.4.1 Outcrop Fracture Observation

Data for some Yingcheng Formation outcrops in the Songliao Basin were collected from previous studies (Fig. 2.17; Wang et al. 2008) to gain a direct insight into fracture dimension, density, and extension for facilitating the analysis of fractures in the target horizon.

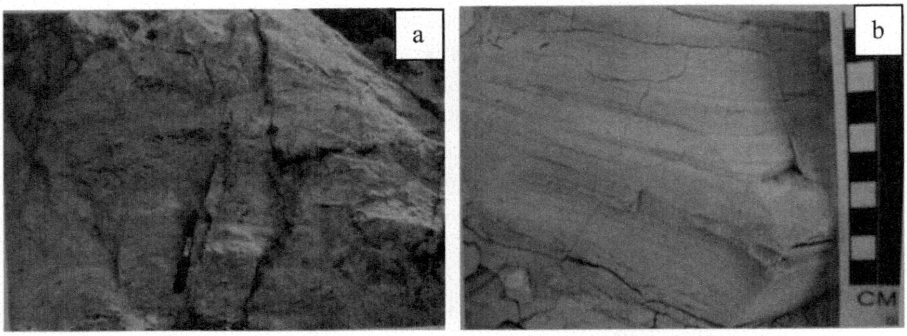

Fig. 2.17 Field outcrop characteristics of volcanic fractures in Yingcheng Formation, Songliao Basin (Wang et al. 2008). **a** Rhyolite structure rhyolite, development of vertical structural fractures; **b** rhyolitic tuff, fracture network development, Jiutai Shanghewan, Jilin

2.3.4.2 Core Structural Fracture Observation

The core observation is a direct method of fracture characterization. Large vertical and high-angle structural fractures are observed on volcanic core samples acquired from the Xudong area in the Songliao Basin, and they vary in density with location and lithology (Fig. 2.18). Large fractures extend over 0.5 m in length, and a 3D fracture network consisting of numerous small fractures (smaller than 0.1 m in extension) is seen locally.

As per intersection and filling of fractures derived from core observation, structural fractures in the study area are qualitatively classified into three phases (Fig. 2.18). Phase-1 fractures are low-angle structural fractures filled with materials and intersected by high-angle structural fractures; late structural fractures usually intersect early structural fractures. Phase-2 fractures are high-angle structural fractures intersecting low-angle structural fractures and are filled with materials; they are filled with calcite veins or intruded by magma, suggesting that they formed before the open or semi-open high-angle structural fractures. Phase-3 fractures are high-angle open structural fractures intersecting low-angle structural fractures. Early fractures are generally filled with calcite veins and feature large widths above 0.01 m and large extension. The late fractures were mostly open and extend over 0.5 m in length. Approximately one-third of fractures are small in dimension and concentrate to form a fracture network. Fractures of the last two phases are found to have greater impacts on reservoir properties.

2.3.4.3 Thin-Section Fracture Observation

Microscopic observation is one of the most immediate means to yield a full image of fractures (Fig. 2.19). Fractures in the target horizon vary in geometry and dimension. There are large fractures with long extension and small fractures related to late diagenesis. Fractures gathering locally may form a 3D fracture network (Fig. 2.19h). These fractures act as the pathways for hydrocarbon migration and transport but may induce water breakthrough during development to impair the production.

Burst fractures (Fig. 2.19a), dissolution fractures (Fig. 2.19b, c), structural fractures (Fig. 2.19d, e), condensing shrinkage fractures (Fig. 2.19f), and stylolitic fractures (Fig. 2.19g) exhibit distinct features on thin sections, as shown in Table 2.3. Owing to data quality and observation range, dissolution fractures are the most common type observed under a microscope followed by burst fractures; other fracture types are seldom observed. The thin sections (Fig. 2.19e) also show 3-phase fractures (see the orientations on the map for fracture strikes), which agree with core observations.

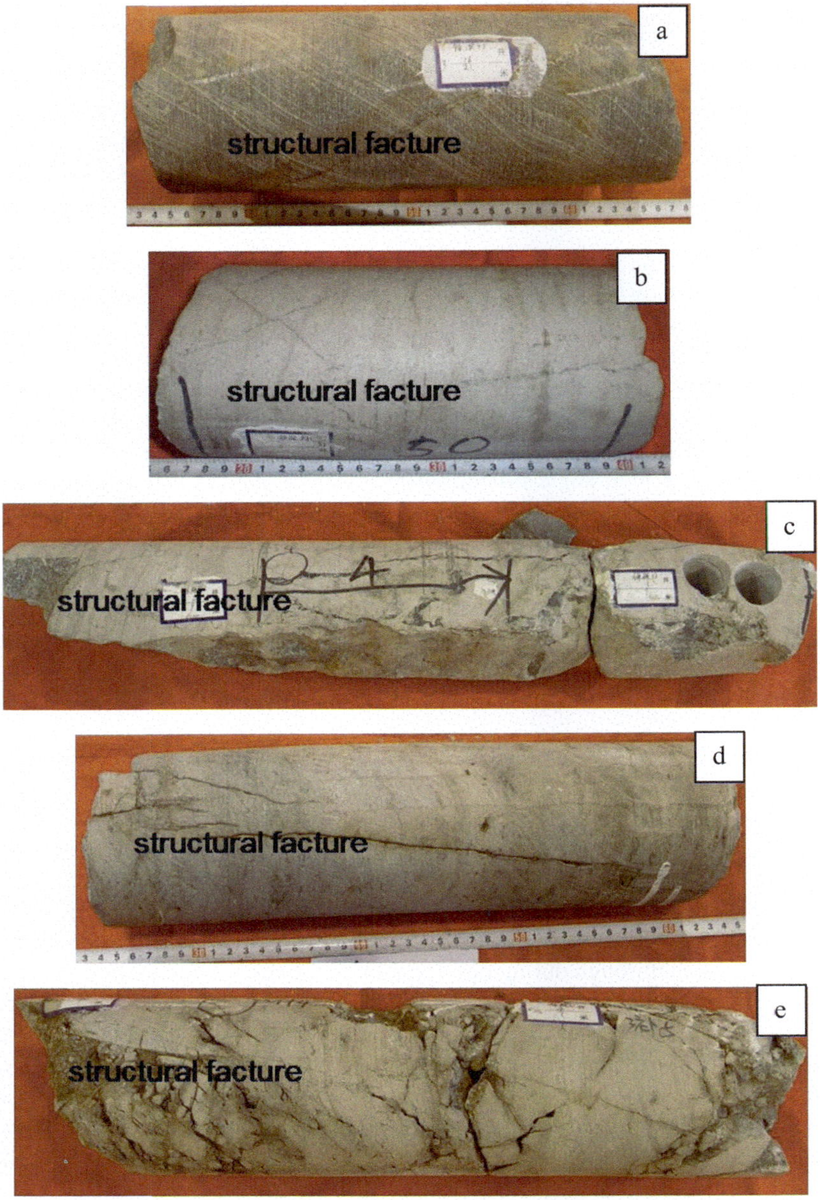

Fig. 2.18 Core observation characteristics of volcanic rock fractures in the first member of Yingcheng Formation in Xudong area, Songliao Basin. **a** High-angle calcite vein filling structural fractures, XS17 well, depth 3649.89–3650.15 m; **b** high-angle calcite vein filling structural fractures, well XS231, depth 3760–3760.23 m; **c** vertical structural fracture, XS12 well, depth 3730.06–3730.57 m; **d** high angle structural fracture, vertical structural fracture, XS21 well, depth 3656.24–3656.58 m; **e** high angle tectonic fracture network, XS12 well, depth 3667.83–3668.24 m

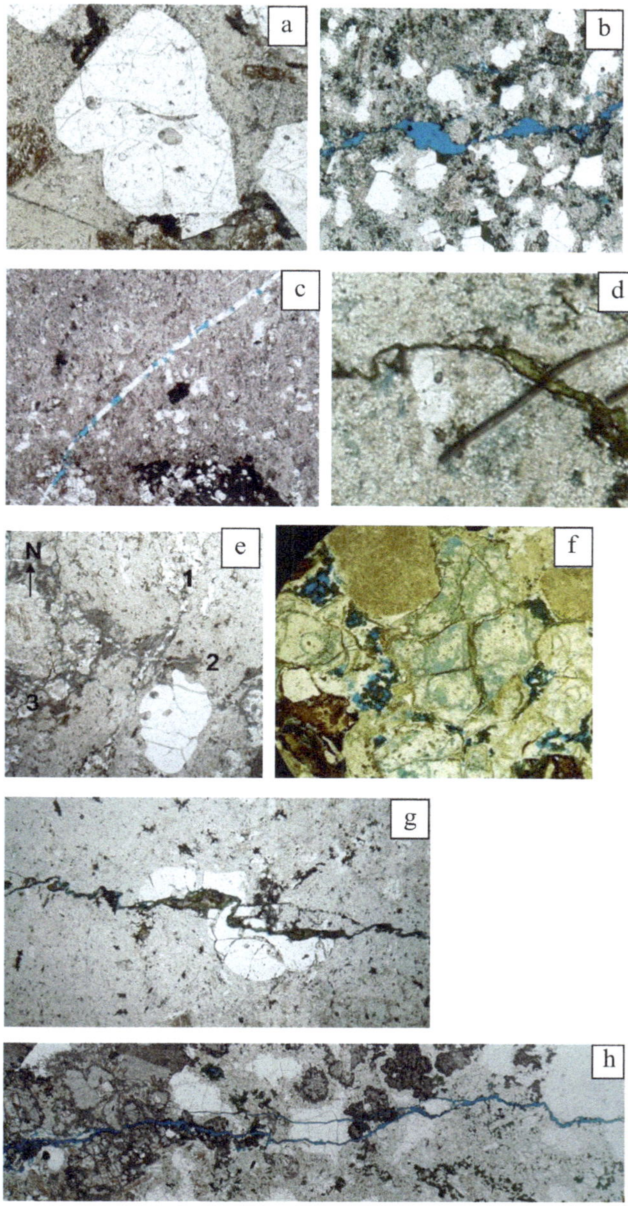

Fig. 2.19 Microscopic thin section characteristics of volcanic rock fractures in the first member of Yingcheng Formation in Xudong area, Songliao Basin. **a** Crystal chip burst fracture, well XS42, depth 3702.06 m, ×10, (−); **b** dissolution fracture, XS 21 well, depth 3658.66 m, ×10, (−); **c** quartz half-filled dissolution fracture, XS 23 well, depth 3899.47 m, ×40, (−); **d** chlorite filling fracture, XS12 well, depth 3731.47 m, ×10, (−); **e** different stages of fracture development characteristics, XS12 well, depth 3611.07 m, ×10, (−); **f** volcanic breccia condensation contraction joint, well XS 401, depth 4176.48 m, ×10, (−); **g** suture, XS 12 well, depth 3731.47 m, ×40, (−) H fracture network, XS21 well, depth 3731.17 m, ×40, (−)

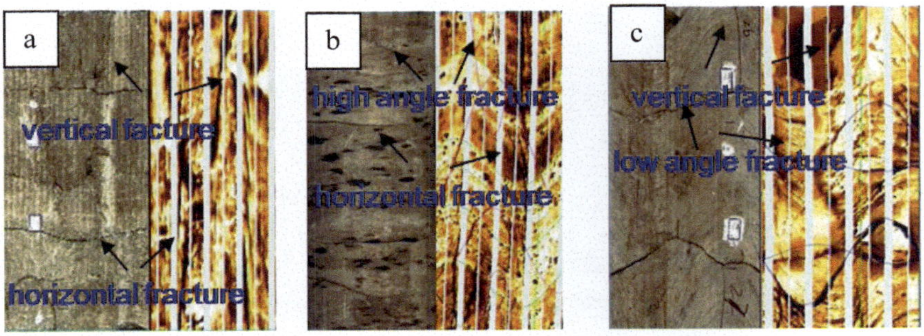

Fig. 2.20 Imaging logging response characteristics of volcanic fractures in the first member of Yingcheng Formation in Xudong area, Songliao Basin. **a** Horizontal fracture and vertical fracture, XS13 well, depth of 4130–4131.2 m; **b** high angle fracture and horizontal fracture, XS14 well, depth 4207.7–4209.3 m; **c** low angle fracture and vertical fracture, XS21 well, depth of 3732.4–3734.4 m

2.3.4.4 Log-Based Fracture Characterization

Some logging techniques can be used for fracture characterization. FMI logging uses electric current and acoustic wave beams to scan the shaft and obtain the "image" of the borehole wall (Wang et al. 1999a, b, c). This technique is useful to fracture characterization. In addition to conventional-logging interpretation of fractures and lithofacies, FMI and cores were combined for a detailed study of fractures in the target horizon (Fig. 2.20). Vertical, high-angle, low-angle, and horizontal fractures, especially phase-3 fractures, show distinct responses in FMI images. Abundant microfractures, usually 1–40 μm wide, can also be imaged on FMI logs.

2.3.4.5 Seismic-Based Fracture Characterization

Coherence analysis was first performed to delineate the lateral distribution of fractures, particularly in the areas with no wells. Figure 2.21a shows coherence-based fracture delineation in the XS27 well field, Xudong area. It can be seen that fractures are most developed in the northeast and southwest but weakly developed in the central area. Figure 2.21b shows the result of ant tracking obtained using Petrel-based on coherence analysis. It can be seen that microfractures mainly turn up along large faults and vary in density from area to area. As compared, ant tracking yields a more detailed image of fractures than coherence analysis because seismic information is fully utilized through the ant tracking algorithm, which is an effective technique for precise fracture characterization in the areas with no wells.

Ant tracking, a module in Petrel, is capable of imaging small faults and fractures in addition to large faults. According to the result of ant tracking, natural fractures in the Xudong area are most developed along the Xuzhong and Xudong faults in a nearly N–S direction. Few large faults are concentrated in the west of the northern and middle parts, while many small faults are scattered evenly in the south of the

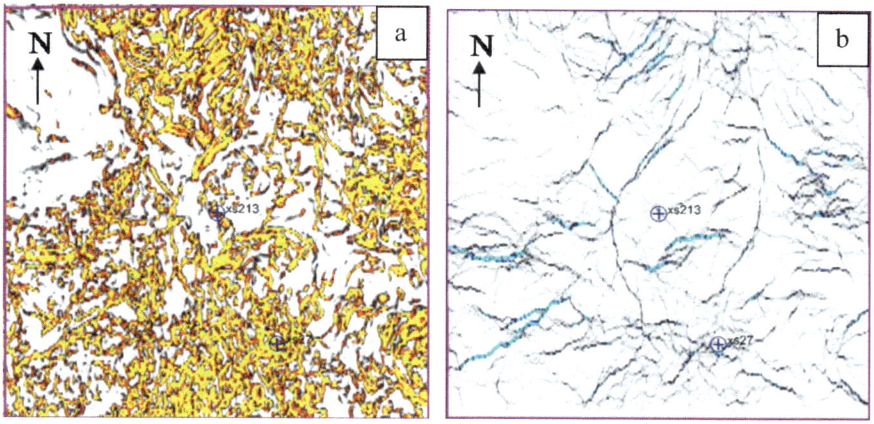

Fig. 2.21 The seismic data of the first member of Yingcheng Formation in XS27 well area of Xudong area in Songliao Basin depict the characteristics of fault development. **a** Seismic coherence analysis of fracture plane characteristics; **b** seismic data ant tracking fracture plane development characteristics. *Note* The dark line in figure B is the development position of the fracture on the plane. The position of XS27 well area is shown in Fig. 2.21

Xudong area (Fig. 2.22). Macroscopically, fractures and faults show a similar trend, and fractures mainly turn up close to faults. Microscopically, in addition to those near-fault fractures in a nearly S–N direction, numerous fractures distribute in the E–W or nearly E–W direction. These fractures connect with S–N fractures, but the lateral extension of the latter is much smaller than that of the former. Various fractures spatially constitute a 3D fracture network in the Xudong area, which is an important control on reservoir properties in the target horizon.

2.3.5 *Fracture Genesis and Controls*

Fractures are formed under the control of various geologic processes, such as local tectonism, regional stress, diagenetic contraction, unloading, weathering, and even sedimentation, which may be varied in different regions (Wu et al. 2005). Fractures in volcanic gas reservoirs in the Ying-1 Member in the Xudong area, the Songliao Basin are diagnosed to be related to tectonism, volcanic lithology, volcanic massif, volcanic facies, and diagenesis. As per spatial fracture distribution and microscopic thin-section features, tectonism and diagenesis are the two major controls followed by volcanic lithology, lithofacies, and massif.

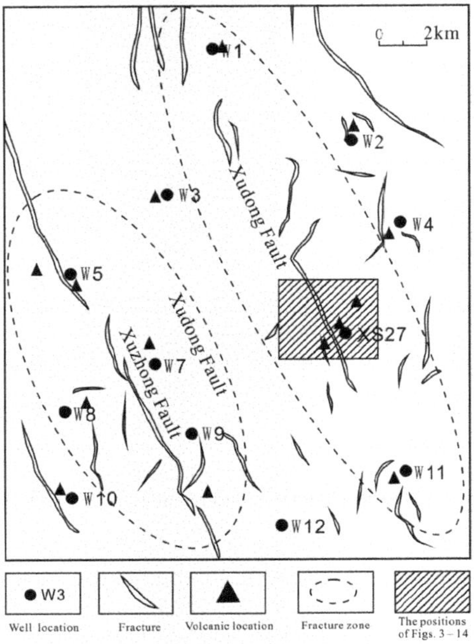

Fig. 2.22 Plane feature map of fault development of volcanic rock strata in the first member of Yingcheng Formation in Xudong area of Songliao Basin

2.3.5.1 Tectonism

Tectonic features are among the major controls in volcanic reservoirs and play a decisive role in the formation of fractures. Tectonism affects fractures from the perspectives of fault activity and ancient landform generated by tectonic uplifting and compaction, especially fault activity. In accordance with the episodes of faulting activities, the fault system in the Xushen gas field is classified into early faults (from the Huoshiling Formation to the Shahezi Formation), inherited faults (from the Huoshiling Formation to the Yingcheng Formation), and late faults (after the Denglouku Formation). Spatially, the fault system consists of two large faults—Xuzhong and Xudong (Fig. 2.22). Previous theoretical studies and observations (Wu et al. 2005) reported that faults and fractures are the products of the same genetic mechanism and that fractures are the embryonic form of faults. Generally, faults are always associated with fractures in the vicinity; they are presumably identical in the stress field.

In view of distribution, laterally, faults and fractures are well-developed around the Xuzhong and Xudong fault zones and underdeveloped far away from these fault zones; vertically, faults and fractures are more developed in the first member of the Yingcheng Formation than the fourth member of the Yingcheng Formation. The Xudong area consists of several faulted anticlines and blocks dominated by volcanic activities and tectogenetic movements. Currently, the study area features structural low in the middle and structural high around. The principal part, that is, the Xudong slope, features structural high in the east and structural low in the west.

Such a landform may affect volcanic massif distribution and consequent formation of fractures.

In view of fissure–central eruption dominating the target horizon, fractures tend to be more developed in the structural highs close to craters than in the structural lows far away from craters. Craters are identified mainly using the seismic data. They feature inverted cone-like reflections and internal chaotic reflections. If the crater is penetrated by a well, lithologic and electric log responses can also be used for identification. As indicated in Fig. 2.22, craters are usually around faults, where there are also many fractures. Figures 2.21 and 2.22 seem to differ because the latter is a more precise local image of the former.

2.3.5.2 Volcanic Lithology

Diverse volcanic rocks in the Ying-1 Member in the Xudong area are the product of multiphase eruption. As per lab tests, microscopic observation, and TAS chart (Fig. 2.23), 2 rock types, that is, volcanic lava and volcaniclastic rocks, and 17 lithologies are detected in the cored interval. The volcanic lava includes acidic rocks, intermediate-acidic rocks, intermediate rocks, and intermediate-basic rocks; majority is the acidic rocks. Rock types include brecciated lava, tuff lava, welded breccia, basaltic brecciated lava, rhyolite, tuff, tuffite, volcanic breccia, basalt, welded tuff, sedimentary volcanic breccia, conglomerate, crystal tuff, basaltic volcanic breccia, tuffaceous breccia, siltstone, and finestone. Rhyolite, brecciated lava, and tuff lava are predominant.

Rhyolites, welded tuffs, and tuffs, lithologically tight, contain structural fractures, while other types of volcanic rocks have fewer fractures. Volcanoclastic rocks such as volcanic breccia, welded breccia, and brecciated lava are dominated by dissolution fractures. The target horizon is mainly composed of acidic rocks (e.g., rhyolites, welded tuffs, and crystal tuffs), with a continuous distribution. Thus, structural fractures in these rocks usually extend over 0.5 m in length and can be easily detected through core observation.

2.3.5.3 Volcanic Massif

There are numerous volcanic massifs in the Ying-1 Member in the Xudong area, the Songliao Basin. Volcanic massifs in different cycles of eruption are spatially superimposed to form the volcanic formations in the Ying-1 Member. Low-angle or horizontal fractures usually reside along the boundaries of the massifs. Massif dimension and the state of superimposition have some impacts on the fracture orientation and spatial extension. Volcanic massifs can be tracked on seismic sections in accordance with the location of the crater and reflection continuity and energy of massifs of different phases. Lithologic and electric log responses can also be used for identification. As indicated by seismic sections, high-angle fractures usually occur close to the volcanic

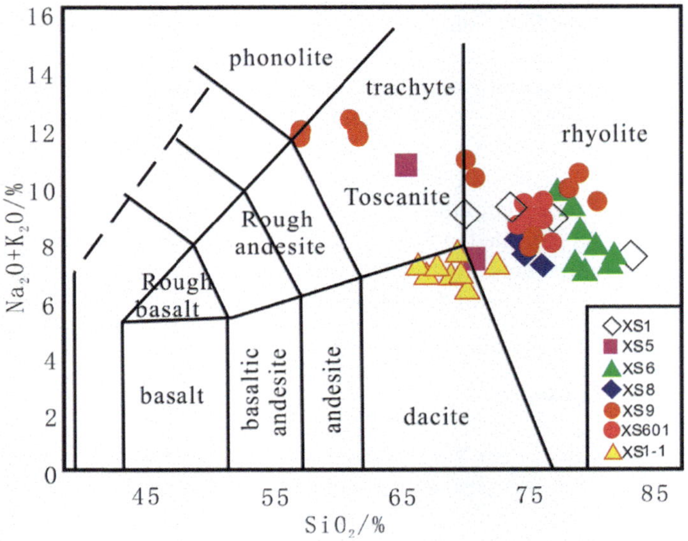

Fig. 2.23 The alkali-silica diagram (TAS diagram) of volcanic rocks in Xudong area, Songliao Basin (Wang et al. 2007a, b, c)

massif boundary around the crater and extend over a long distance vertically. Horizontal and low-angle fractures usually turn at the boundary of the massif far away from the crater and extend over a long distance laterally. These fractures exhibit a belt-shaped geometry with internal chaotic reflections on seismic sections. The fault where seismic events are completely disconnected is usually associated with fractures (Fig. 2.24). Fracture interpretation should be performed on several neighboring or perpendicular seismic sections to alleviate uncertainty. Fracture dimension and dip angle are also dependent on the landform to some extent. Laterally, fractures mainly turn up in the area of the massif close to the crater; fracture density decreases with increasing distance away from the crater.

2.3.5.4 Volcanic Facies

Volcanic rocks in the target horizon are classified into 5 facies and 16 subfacies (Table 2.4). The volcanic conduit facies of 30-m thick features high-magnitude dentiform log responses. Extrusive facies generally below 20-m thick is close to volcanic conduit facies and features medium- to low-magnitude log responses. Explosive facies shows medium to low dentiform resistivity log. Effusive facies of large thickness shows slightly dentiform medium to high resistivity log. Volcanogenic sedimentary facies of varied thickness commonly features rhythmic log responses. In accordance with the relations between log-interpreted fracture porosity/fracture permeability and volcanic facies, volcanogenic sedimentary facies shows the best

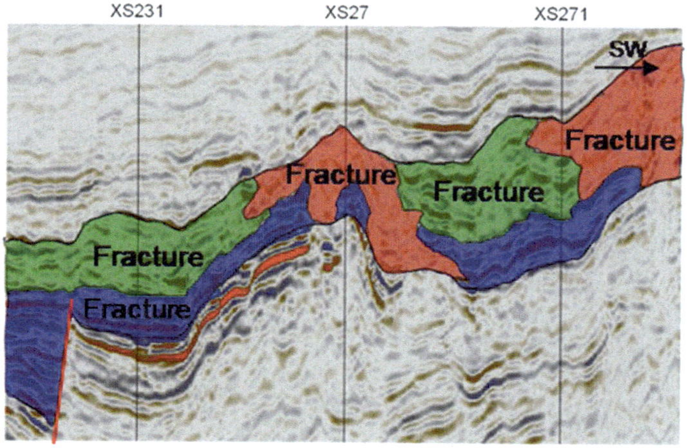

Fig. 2.24 The well-seismic data of the first member of Yingcheng Formation in Xudong area of Songliao Basin combined with the identification characteristics of volcanic rock mass. *Note:* Different color areas in the figure are different volcanic rock bodies

fracture petrophysical properties and extrusive facies shows the poorest properties; other volcanic facies lie between them (Fig. 2.25). Volcanogenic sedimentary facies is generally composed of various volcaniclastic rocks with poor intergranular cementation, in which fractures may easily form. Fault-fracture distribution and crater-fracture correlation indicate fissure–central eruption in the study area, which indicates more developed fractures in volcanic conduit facies and explosive facies and less developed fractures in effusive facies and volcanogenic sedimentary facies far away from craters. The more developed fractures in volcanogenic sedimentary facies than in effusive facies are mainly attributed to lithology. Fracture index is defined using fracture porosity and fracture permeability as follows to characterize the degree of fracture development (Kuang et al. 2001).

$$F = \phi_f \cdot K_f \cdot h \times 100 \tag{2.1}$$

where ϕ_f is the fracture porosity, %; K_f is the fracture permeability, mD; and h is the thickness of the reservoir, m.

As indicated by the fracture index and the histogram of fracture width (Fig. 2.26), volcanic conduit facies shows the most developed fractures. The fracture index can be derived from the fracture porosity and fracture permeability, and the fracture width is calculated using FMI as follows:

$$\varepsilon = aAR_{xo}^{b}R_{m}^{1-b} \tag{2.2}$$

where a and b are the constants related to the logging tool, and b is close to zero; A is the area with conductivity anomaly caused by the existence of fractures, mm^2; and

Table 2.4 Volcanic facies in the Ying-1 Member in the Xudong area, the Songliao Basin

Facies	Per-well thickness percentage (%)	Subfacies
Volcanic conduit facies	4.55	Volcanic neck, subvolcano, and cryptoexplosive breccia
Extrusive facies	1.35	Intrazone, mesozone, and exozone
Explosive facies	44.71	Avalanche, thermal pyroclastic flow, thermal base surge, and fallout
Effusive facies	43.40	Top, upper, middle, and lower
Volcanogenic sedimentary facies	5.99	Epiclast-bearing, and retransported

R_{xo} and R_m are the resistivities of the invaded zone and drilling fluid, respectively, Ω m.

All the data can be derived from well logging for the calculation of fracture width.

2.3.5.5 Diagenesis

From the perspective of oil and gas reservoirs, volcanic diagenesis is defined as the sum of processes of eruptive material, that is, magmatic melts, and/or volcaniclastic material changing into rocks before the conversion into metamorphic rocks or weathering product. Diagenetic effects on fractures, which can be detected through thin-section observation, may be positive (e.g., condensing contraction, corrosion, and weathering) or negative (e.g., compaction and packing) (Fig. 2.20). The condensing contraction is related to the condensing contraction of the eruptive material. Corrosion is classified into partial intragranular corrosion (intracrystal fragment corrosion and interdebris corrosion of volcanic breccia), complete granular corrosion to form moldic pores (feldspar phenocrysts or rock debris completely corroded to form moldic pores), pyromeride matrix corrosion, and intergranular corrosion. Intergranular corrosion has a stronger impact on fractures than intragranular corrosion. Weathering refers to various mechanical or chemical processes on rocks exposed to surface or near-surface conditions, and it forms weathered fractures to improve reservoir properties. Weathering has less effect on fractures than condensing contraction and corrosion in the target horizon. Compaction refers to the process of the volcaniclastic material changing into rocks through compacted consolidation. Through microscopic observation, it is believed that the target horizon has experienced packing processes such as albite packing, authigenic quartz packing, chlorite packing, and carbonate packing. Chlorite packing has the largest negative impact on fractures. Compared with packing, compaction is much less destructive to fractures. Corrosion and packing are the most common diagenetic processes in the study area.

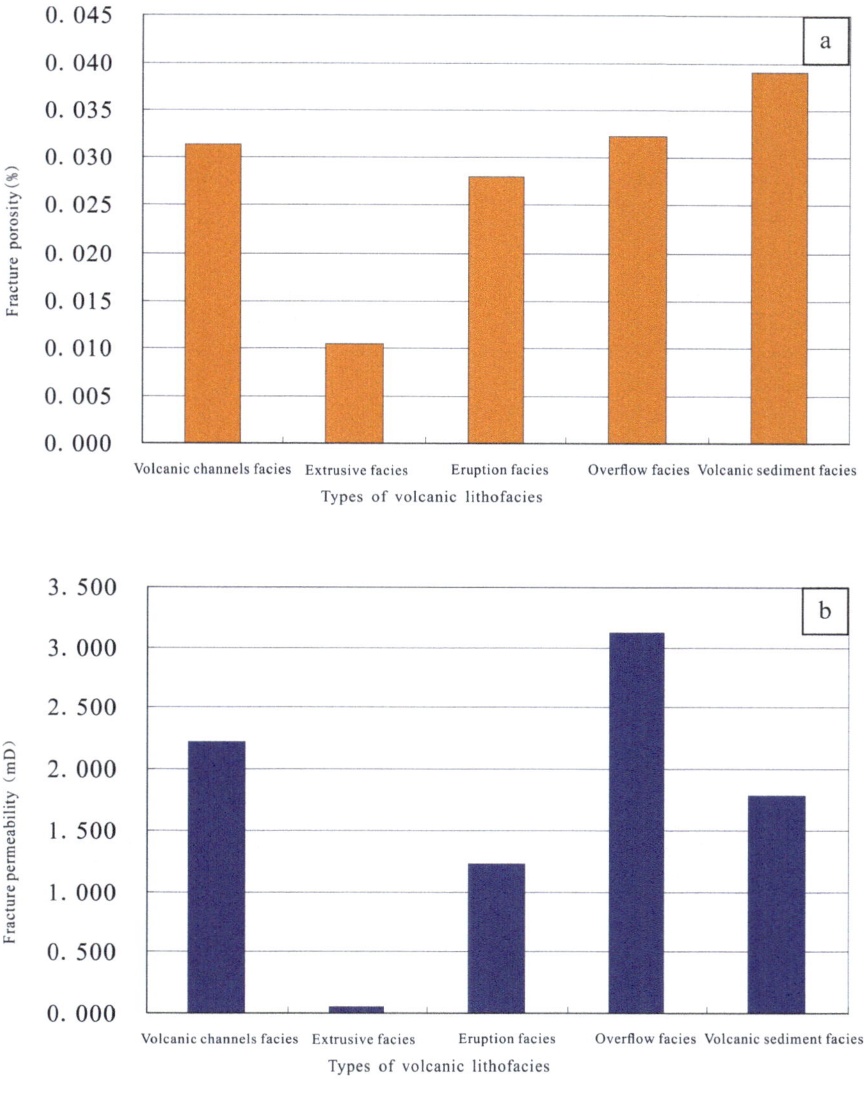

Fig. 2.25 The relationship between reservoir fracture properties and volcanic facies interpreted by single well in the first member of Yingcheng Formation in Xudong area, Songliao Basin **a** single well logging interpretation of reservoir fracture porosity **b** single well logging interpretation of reservoir fracture permeability

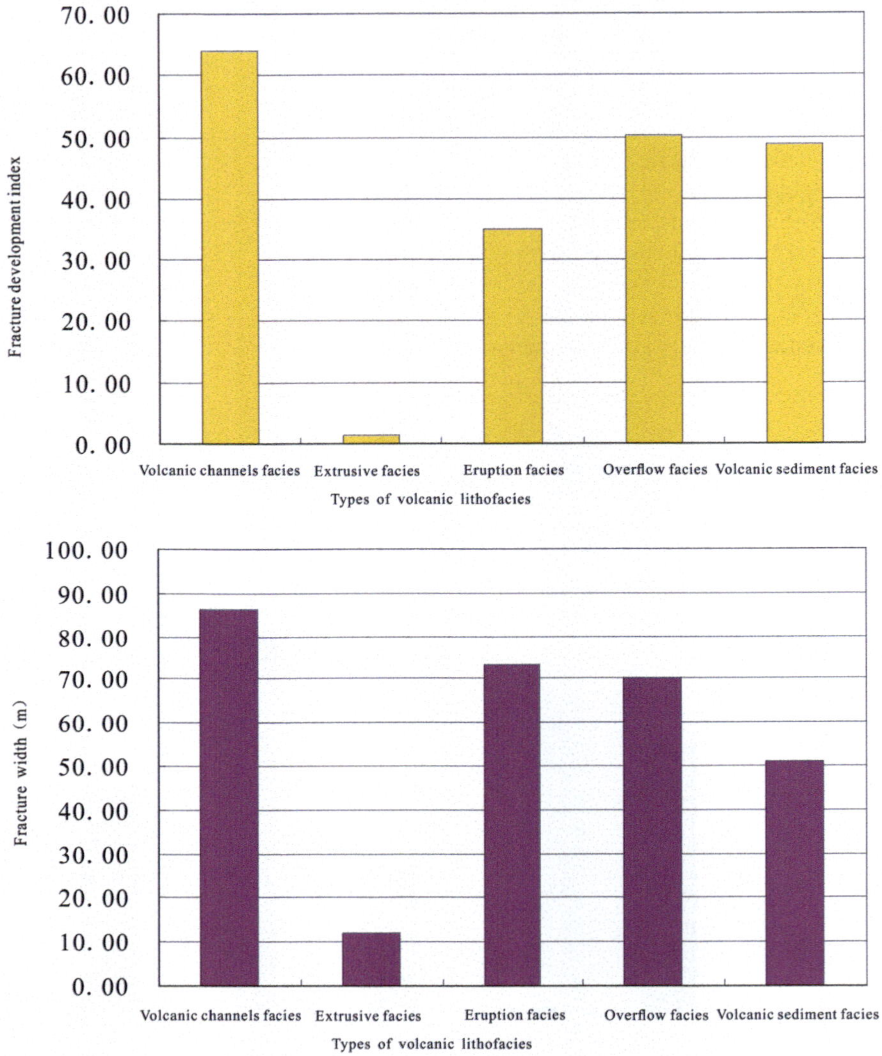

Fig. 2.26 Relationship between fracture development degree and volcanic facies of the first member of Yingcheng Formation in Xudong area, Songliao Basin **a** single well logging interpretation of reservoir fracture development index **b** single well logging interpretation of reservoir fracture width

It is worth noting that fractures in volcanic gas reservoirs in the Ying-1 Member in the Xudong area, the Songliao Basin originate from the interaction of controls above. For example, fractured unconsolidated rocks in an area rich in structural fractures tend to corrode to form dissolution fractures (Fig. 2.19e). The birth of a fracture may be attributed to several factors despite their different contributions. Figure 2.19a shows the fractures as the product of packing evolving from the tectonism-induced fractures. Thus, comprehensive analysis should be performed to identify the genetic mechanism of fractures.

2.4 Summary

Geological genetic analysis of structures plays an important role in oil/gas field development. In other countries, this research has made great progress considering solid minerals; various techniques, including basic geology (e.g., outcrop observation), petrology, and geological modeling, have been used to investigate reservoirs of multiple genetic types, such as clastic, carbonate, volcanic, and metamorphic rocks; however, the geological genetic analysis of structures is rarely performed for petroleum exploration and development, seldom microscopic, and insufficiently based on well logging data. In China, relevant research has involved petroleum exploration and development with some achievements obtained, mainly by virtue of basic geology, well-seismic interpretation, and physical simulation of tectonic origin; considerable attention is paid to the geological genesis of low-relief structures; nonetheless, the geological genetic analysis of structures is not adequately integrated with sedimentology and diagenesis, many problems exist in geological modeling, and the research subjects are mainly limited to clastic rocks.

Geological genetic analysis of structures in field development involves eight aspects—(1) structural evolution history and in situ stress field are investigated to determine the geological genesis of structures; (2) fine structure interpretation and fault grouping are performed using well and seismic data to delineate structures and analyze their geological genesis; (3) methods for geological genetic analysis of structures, including outcrop observation, logging interpretation, seismic interpretation, lab tests, physical simulations and numerical simulations, are improved; (4) geological genetic analysis of reservoirs is conducted by combining dynamic and static data to determine major controls on structural genesis; (5) geological genetic analysis of structures is combined with studies on sedimentary facies and diagenesis; (6) the results of geological genetic analysis are actively applied to field development; (7) fracture characterization and modeling are conducted; and (8) geological genesis of microstructures is analyzed. The techniques for geological genetic analysis of structures in field development mainly include basic geologic analysis, outcrop observation and description, core observation and description, thin-section observation, physical simulations, well logging interpretation, seismic interpretation, laboratory tests and statistics, geological modeling, and dynamic monitoring and production performance analysis.

There are some problems in geological genetic analysis of structures—(1) research findings are biased by insufficient with sedimentological and diagenetic studies; (2) qualitative investigation based on geologic and geophysical techniques overwhelms quantitative investigation; (3) geological genetic analysis often relies on and is restricted to a simple method that has its limitations; (4) geological modeling, especially of fractures, leaves much to be desired, with uncertainty in the accuracy of spatial prediction; (5) existing techniques for geological genetic analysis of structures are not adequate and should be improved along with related disciplines; (6) macroscopic studies overwhelm microscopic studies in geological genetic analysis of structures; (7) geological genetic analysis of structures mainly focuses on clastic rocks and seldom on carbonate, volcanic and metamorphic rocks; and (8) geological genetic analysis of structures needs to be expanded to show its significance and value. The geological genetic analysis of structures is believed to further advance in eight prospects—(1) it is combined with sedimentological and diagenetic studies for comprehensive diagnosis of tectonic origin; (2) more quantitative efforts are taken for geological genetic analysis of structures; (3) certain techniques are integrated for a comprehensive geological genetic analysis of structures to achieve optimal results; (4) the geological modeling algorithm is optimized to expand the technique's application in geological genetic analysis of structures; (5) new techniques and methods are figured out for geological genetic analysis of structures; (6) microstructures and fractures are incorporated to improve the accuracy of geological genetic analysis of structures; (7) geological genetic analysis of structures is extended to reservoirs of other lithologies such as carbonate, volcanic, and metamorphic rocks; and (8) geological genetic analysis of structures is gradually promoted to other works.

Detailed fault analysis is important for oil/gas field development and EOR. Fault analysis in field development depends on two foundations, that is, data and techniques. Relevant data mainly include basic geology, outcrop, well drilling, seismic, dynamic monitoring, and production performance data. Relevant techniques primarily include basic geological analysis, outcrop observation and description, core observation and description, thin-section observation, physical simulations, well logging interpretation, seismic interpretation, laboratory tests and statistics, geological modeling, and dynamic monitoring and production performance analysis.

The location, dimension, and the distribution of faults and fractures, reservoir architecture, fine fault interpretation and modeling, and fault sealing are closely related to efficient field development and should be considered seriously. Faults and fractures dominate reservoir properties and remaining oil distribution, and fracture architecture and fault sealing affect the connection between injectors and producers. Thus, fault modeling is important for interwell reservoir prediction.

Fault analysis is performed to develop various lithologic reservoirs, including sandstone, glutenite, volcanic rock, and carbonatite, and covers fault system genesis, spatial distribution and dimension of faults and fractures, fault sealing, fracture influence on reservoir properties, multi-information fracture characterization, and 3D geological modeling of fractures. Specifically, the current challenges and hotspots in field development within China are detailed as characterization of natural and dynamic fractures in low-permeability reservoirs of the Changqing Oilfield, accurate fault interpretation and delineation using the data of widely spaced wells and seismic data for deep reservoirs in the Tarim Basin, and integration of well and seismic data with dynamic monitoring and production performance data for detailed interpretation and delineation of fourth- and lower-order faults in the Daqing Oilfield.

In view of genesis, fractures in volcanic reservoirs in the Ying-1 Member in the Xudong area, the Songliao Basin are classified into structural fractures, condensing shrinkage fractures, burst fractures, dissolution fractures, stylolitic fractures, and weathered fractures. Structural fractures and dissolution fractures are the most developed types.

Structural fractures are easily observed on volcanic core samples acquired from the Ying-1 Member in the Xudong area, the Songliao Basin, most of which are high-angle or vertical, with the extension generally over 0.5 m. Dissolution fractures are mainly detected under a microscope. Fractures in the target horizon are classified into three phases. Fine fracture characterization is accomplished using core, thin-section, logging, FMI, and seismic data. Ant tracking, a module in Petrel, yields a more detailed image of fractures in the target horizon than coherence analysis.

As indicated by ant tracking, fractures in volcanic reservoirs in the Ying-1 Member in the Xudong area, the Songliao Basin are most developed along the Xuzhong and Xudong faults, roughly in a nearly N–S direction. A small number of large faults concentrate in the northern and central parts of the study area, and many small faults scatter in the southern part. Fractures and faults show a similar trend, and fractures mainly turn up close to faults. In addition to those near-fault fractures in an S–N direction, numerous fractures distribute in the E–W or nearly E–W direction. These fractures connect with S–N fractures, but the lateral extension of the latter is much smaller than that of the former.

References

Ameen MS, Buhidma IM, Rahim Z. The function of fractures and in-situ stresses in the Khuff reservoir performance, onshore fields, Saudi Arabia. AAPG Bull. 2010;94(1):27–60.

Botros NS. The role of the granite emplacement and structural setting on the genesis of gold mineralization in Egypt. Ore Geol Rev. 2015;70:173–87.

Braathen A, Jan T, Haakon F, et al. Fault facies and its application to sandstone reservoirs. AAPG Bull 2009;93(7):891–917.

Casas-Sainz AM, de Vicente G. On the tectonic origin of Iberian topography. Tectonophysics. 2009;474:214–35.

Charvet J, Shu L, Faure M, et al. Structural development of the Lower Paleozoic belt of South China: Genesisof an intracontinental orogen. J Asian Earth Sci. 2010;39:309–30.

Chen F, Luo P, Zhang X, et al. Characteristics of grain size of delta sandbody framework in Yanchang Formation Upper Triassic, North Shaanxi. Acta Sediementologica Sinca. 2010a;28(1):58–67.

Chen H, Zhu X, Zhang G, et al. Characteristics of migration system of Paleogene Lingshui Formation in the deep water area in Qiongdongnan Basin. Acta Geol Sin. 2010b;84(1):138–48.

Chen H, Hu Y, Ran Q, et al. Lithofacies characteristics of volcanic gas reservoirs and their influence on reservoir physical properties:a ease study of Member 1 of Cretaceous Yingcheng Formation in Xudong area of the Xushen gas field. Acta Petrologica Et Mineralogica. 2011a;30(1):60–70.

Chen H, Hu Y, Jin J, et al. Multiple sets of information synthesized to describe fractures of volcanic reservoir: taking volcanic reservoir of the Member 1 of Yingcheng Formation in Xudong Area of Xushen Gas Field as an example. Earth Sci Front. 2011b;18(2):294–303.

Chen H, Hu Y, Jin J, et al. Researches on flow unit of the volcanic reservoir in 1st Member of lower Cretaceous Yingcheng formation, Xudong area, Songliao Basin. Geol China. 2011c;38(6):1430–9.
Chen H, Zhu X, Zhang G, et al. Seismic facies in a deepwater area of a marine faulted basin: deepwater area of the Paleogene Lingshui Formation in the Qingdongnan Basin. Acta Geologica Sinca. 2012a;86(2):473–83.
Chen H, Xu L, Liu C, et al. Classified evaluation of the geological bodies in Yulou Oil Reservoirs of LiaoheBasin. Petrol Geol Oilfield Dev Daqing. 2015a;34(5):45–51.
Chen H, Liang S, Jian P, et al. Flow barrier characteristics of thermal recovery of heavy oil reservoir: taking Yulou oil-bearing sets in some experimental area in the south end of westward slope of West depression in Liaohe Basin as an example. Acta Sedimentol Sin. 2015b;33(3):616–24.
Chen H, Liang S, Shu Z, et al. Characteristics of conglomerate reservoir architecture of alluvial fan and its controlling effects to reservoir development: taking alluvial fan reservoir in some area of Northwest Margin of Junggar Basin as an example. J Jilin Univ (Earte Sci Edn). 2015c;45(1):13–24.
Chen H, Shi C, Cao C. Problems with the fine description of reservoirs. Pet Geol Exp. 2016a;38(5):569–76.
Chen H, Hu Y, Shi C. Single factor characterization and quantitative evaluation of volcanic rock reservoirs in songliao basin. Petroleum Industry Press; 2016b.
Chen H, Hu H, Wu H, et al. Geological genetic analysis of oil and gas reservoir development. Chinese J Geol. 2019a;54(1):192–212.
Chen H, Liang S, Li W. Advances of genesis analysis on structural geology in oil and gas field development. Geol J China Univ. 2019b;25(3):444–56.
Chen P, Wang Q, Guo L, et al. Multi-parameters quantitative evaluation for carbonate reservoir based on geological genesis. J Southwest Petrol Univ (Sci Technol Edn). 2019c;41(4):55–64.
Chen H, Hu Y, Ran Q, et al. Mode characteristics of volcanic eruption of the Member 1 of Yingcheng Formation in Xudong Area in Songliao Basin. J Southwest Petrol Univ (Sci Technol Edn). 2012;34(1):41–48.
Cooperdock EHG, Raia NH, Barnes JD, et al. Tectonic origin of serpentinites on Syros, Greece: geochemical signatures of abyssal origin preserved in a HP/LT subduction complex. Lithos. 2018;296–299:352–364.
Corfield RI, Searle MP, Pedersen RB. Tectonic setting, origin, and obduction history of the spontang ophiolite, Ladakh Himalaya, NW India. J Geol. 2001;109:715–36.
Corver MP, Doust H, van Wees JD. Classification of rifted sedimentary basins of the Pannonian Basin System according to the structural genesis, evolutionary history and hydrocarbon maturation zones. Marine Petrol Geol. 2009;26:1452–1464.
d'Huteau E, Repsol YPF, Breda E, et al. Stimulation with hydraulic fracture of an upper cretaceous fissured tuff system in the San Jorge Basin, Argentina. SPE69586; 2001:1–11.
Dai J, Xu J, Meng Z, et al. Applied of limited become deformed methods in volcanic rock cracks forecasting. J Univ Petrol China 2003; 27(1):1–3, 10.
Deibert JE, Camilleri PA. Sedimentologic and tectonic origin of an incised-valley-fill sequence along an extensional marginal-lacustrine system in the Basin and Range province, United States: implications for predictive models of the location of incised valleys. AAPG Bull. 2006;90(2):209–35.
Deng H, Wang H, Zhu Y, et al. High-resolution sequence stratigraphy-principle and application. Beijing: Geological Publishing House; 2002.
Dong D, Xiang L, Cheng J, et al. Stress-mechanism of Qudi Faults Nose and its controlling on tectonic structure and oil distribution. Geol Technol Inform. 2017;36(3):27–32.
Dufréchou G, Harris LB. Tectonic models for the origin of regional transverse structures in the Grenville Province of SW Quebec interpreted from regional gravity. J Geodyn. 2013;64:15–39.
Feng C, Zhang P, Qi B, et al. Recent tectonic stress field at the Shallow Earth's crust near the Tan-lu Fault Zone. Geoscience. 2017;31(1):46–70.

Ge J, Zhang Z, Ni J, et al. Features and dynamic mechanism of the Southern End of Muping-Jimo Fault. Geoscience. 2015;29(4):747–53.
Gou Q, Xu S, Hao F, et al. Study on characterization of micro-fracture of shale based on micro-CT. Acta Geol Sin. 2019;93(9):2372–82.
Han D. Discussions on concepts, countermeasures and technical routes for the redevelopment of high water-cut oilfields. Pet Explor Dev. 2010;37(5):583–91.
Hofmann A, Dirks PHGM, Jelsma HA, et al. A tectonic origin for ironstone horizons in the Zimbabwe craton and their significance for greenstone belt geology. J Geol Soc Lond. 2003;160:83–97.
Hu Q, Dawei D, Li Z, et al. Tectonic evolutionary characteristics and their causes of Chepaizi Uplift in Junggar Basin. Oil Gas Geol. 2016;37(4):556–64.
Ibouh H, Michard A, Charrie A, et al. Tectonic-karstic origin of the alleged "impact crater" of Lake Isli (Imilchil district, High Atlas, Morocco). C. R. Geoscience. 2014;346:82–89.
James NP, Bone Y, Carter RM, et al. Origin of the Late Neogene Roe Plains and their calcarenite veneer: implications for sedimentology and tectonics in the Great Australian Bight. Aust J Earth Sci. 2006;53:407–19.
Jime′nez-Milla′n J, Nieto LM. Geochemical and mineralogical evidence of tectonic and sedimentary factors controlling the origin of ferromanganese crusts associated to stratigraphic discontinuities (Betic Cordilleras, SE of Spain). Chem Erde. 2008;68:323–36.
Klimchouk A, Auler AS, Bezerra FHR, et al. Hypogenic origin, geologic controls and functional organization of a giant cave system in Precambrian carbonates, Brazil. Geomorphology. 2016;253:385–405.
Kuang J, Xu G, Wang Y, et al. The modeling for single well of prediction of fractures and production capacity in compact clastic reservoir rock—example for Saximiao formation in XZ gasfield. J Miner Petrol. 2001;21(2):62–67.
Lavenu APC, Lamarche J, Gallois A, et al. Tectonic versus diagenetic origin of fractures in a naturally fractured carbonate reservoir analog (Nerthe anticline, southeastern France). AAPG Bull. 2013;97(12):2207–32.
Li Y. Structural characterization of complex fault block reservoir. Beijing: Petroleum Industry Press; 2001.
Li Y, Liu J. Reservoir development geology. Beijing: Petroleum Industry Press; 2007. p. 34–35.
Li C, Li L, Li W. The mechanism of volcanic rock fracture formation of Eastern Sag of Liaohe Depression. J Jilin Univ (Earth Sci Edn). 2004a;34(5):46–50 (in Chinese).
Li Q, Simo JA, McGowran B, et al. The eustatic and tectonic origin of Neogene unconformities from the Great Australian Bight. Mar Geol. 2004b;203:57–81.
Li X, Wang P, Hu J, et al. Microstructure characteristic and generation analysis in Gaoji Oilfield. Coplex Hydrocarbon Reserv. 2009a;2(2):16–9.
Li Y, Song Z, Li Q, et al. Prediction of distribution characterss of volcanic facies by using seismic and logging facies. J Mineral Petrol. 2009b;29(1):106–13.
Li X, Yao J, Liu H, et al. Tectonic origin type of the low amplitude uplifted structure and its role in the control of the distribution of oil and gas in Mesozoic, the Ordos Basin. Geoscience. 2013;27(4):755–764, 773.
Li H, Liu S, Wei X, et al. Method exploration of implicit logging geological information research. Prog Geophys. 2015a;30(1):195–202.
Li W, Ren J, Liu Y, et al. Development and formation mechanism of Cenozoic tectonics in the East Area of Liaodong Bay Depression. Geol Sci Technol Inform. 2015b;34(6):58–63.
Liu Z, Xin Q, Deng J, et al. Formation mechanism and structural model of fault block group oil and gas reservoirs. Beijing: Petroleum Industry Press; 1998a.
Liu H, Wu S, Pin Y. Fracture research the latest progress and research advances in the fracture of the South China sea. Nanhai Yanjiu Yu Kaifa. 1998b;35–42.
Liu J, Li S, Wang H. Application of sequential indicator random simulation method to study sedimentary microfacies. Petrol Geol Oilfield Dev Daqing. 2003a;22(2):11–46.

Liu L, Xie W, Jiao L, et al. Discussed of volcanic rock cracks formation of Cenozoic in the East Sag in Liaohe Fault-down Basin. Special Oil Gas Reserv. 2003b;10(1):18–21.

Liu Z, Liu H, Dai S, et al. Study on volcanic rock fracture quantitative identification method using logging data. Petrol Geol Dev Daqing. 2008;27(5):132–4.

Liu X, Wang J, Liu Y, et al. Design and implementation of the delta reservoir geological knowledge base system. Beijing: Petroleum Industry Press; 2014a.

Liu Y, Chen Q, Hu K, et al. Comparison of the Bohai Bay Basin and Subei-South Yellow Sea Basin in the structural characteristics and forming mechanism. Geotecton Metallog. 2014b;38(1):38–51.

Luo Q, Jiang Z, Pang X. Reservoir-controlling mechanism and model of fracture. Beijing: Petroleum Industry Press; 2007.

Lv Y, Fu G, Zhang Y, et al. Study on fault sealing. Beijing: Petroleum Industry Press; 2002.

Lv Y, Fu G, Fu X, et al. The grooming and sealing effect of fault on oil and gas.Beijing: Petroleum Industry Press; 2013.

Martín-Izard A, Gumiel P, Arias M, et al. Fuertes-Fuente, R. Reguilon. Genesis and evolution of the structurally controlled vein mineralization (Sb-Hg) in the Escarlati deposit (León, Spain): Evidence from fault population analysis methods, fluid-inclusion research and stable isotope data. J Geochem Explor. 2009;100:51–66.

Mcmechan M. Nature, origin and tectonic significance of anomalous transverse structures, southeastern Skeena Fold Belt, British Columbia. Bull Can Petrol Geol. 2007;55(4):262–74.

Morteani G, Kostitsyn Y, Preinfalk C, Gilg HA. The genesis of the amethyst geodes at Artigas (Uruguay) and the paleohydrology of the Guarani aquifer: structural, geochemical, oxygen, carbon, strontium isotope and fluid inclusion study. Int J Earth Sci (Geol Rundsch). 2010;99:927–947.

Mualla Cengiz C, Orbay N. The origin of Neogene tectonic rotations in the Galatean volcanic massif, central Anatolia. Int J Earth Sci (Geol Rundsch). 2010;99:413–426.

Neill I, Gibbs JA, Hastie AR, et al. Origin of the volcanic complexes of LaDésirade, Lesser Antilles: implications for tectonic reconstruction of the Late Jurassic to Cretaceous Pacific-proto Caribbean margin. Lithos. 2010;120:407–20.

Nie K, Lu Z, Wang H, et al. Rock flexural strength methods used in volcanic cracks reservoir forecasting. Oil Geophys Prospect. 2007;27(8):31–4.

Pierre FD, Festa A, Irace A (2007) Interaction of tectonic, sedimentary, and diapiric processes in the origin of chaotic sediments: an example from the Messinian of Torino Hill (Tertiary Piedmont Basin, northwestern Italy). GSA Bull. 2007;119(9/10):1107–1119.

Rothery E. Tectonic origin of the shape of the Broken Hill lodes supported by their structural setting in a high-grade shear zone. Aust J Earth Sci. 2001;48:201–20.

Saalmann K, Manttari I, Nyakecho C, et al. Age, tectonic evolution and origin of the Aswa Shear Zone in Uganda: Activation of an oblique ramp during convergence in the East African Orogen. J African Earth Sci. 2016;117:303–330.

Sàbat F, Gelabert B, Rodríguez-Perea A, et al. Geological structure and evolution of Majorca: implications for the origin of the Western Mediterranean. Tectonophysics. 2011;510:217–38.

Shang X, Long S, Duan T, et al. Current situation and development trend of fracture characterization and modeling techniques in shale gas reservoirs. Nat Gas Geosci. 2021;32(2):215–232.

Shen C, Li X, Du X, et al. Some applications in the study of fluid inclusions in oilfield fracture. Petrol Geol Oilfield Dev Daqing. 2003;22(4):4–6.

Stenger BA, Ameen MS, Sa'ad Al-Q, et al. Pore pressure control of fracture reactivation in the Ghawar field, Saudi Arabia. 2002;77642:1–11.

Sun W, Fang M, Li W, et al. The orderliness of fault s and three stages of fault researches. Geological. 2000;305–311.

Tang X, Wang X, Zhu Y. Cracks development level evaluated by synthetical probability methods. Gas Explor Dev. 2009;32(1):26–27, 38.

van Leeuwen T, Allen CM, Kadarusman A, et al. Petrologic, isotopic, and radiometric age constraints on the origin and tectonic history of the Malino Metamorphic Complex, NW Sulawesi, Indonesia. J Asian Earth Sci. 2007;29:751–77.

Wall GRT, Jenkyns HC. The age, origin and tectonic significance of Mesozoic sediment-filled fissures in the Mendip Hills (SW England): implications for extension models and Jurassic sea-level curves. Geol Mag. 2004;141(4):471–504.

Wang Z. Fractures reservoir description and prospecting. Beijing: Petroleun Industry Press; 1999a.

Wang N. Summary of reservoir development model in China. Petroleum Industry Press; 1999c.

Wang Y, Hu Y, Ran Q, et al. Lithology distinguishing technology research of deep layer volcanic gas reservoir in Xushen Gas Field. J Southwest Petrol Univ. 2007a;29(2):78–81.

Wang Z, Ou C, Li F. Study on the lithology discrimination methods for volcanic reservoirs. World Well Log Technol. 2007b;22(1):8–11.

Wang Y, Hu Y, Ran Q, et al. Deep layer cracks developed level evaluated of olcanic gas reservoir. Nat Gas Ind. 2007c;27(8):31–34.

Wang J, Zhang S. Exploring the characteristics and genesis of low amplitude structures on the Yishaan Slope, OrdosBasin.Earth Sci Front. 2018;25(2):246–253.

Wang P, Feng Z, et al. Basin volcanic lithology, lithofacies, reservoir, Gas accumulation and prospecting. Beijing: Press of Science; 2008.

Wang G, Yang S, Liao F, et al. Hierarchical Structure of Barrier beds and Interbeds in Braided River Reservoirs. Nat Gas Geosci. 2009;20(3):278–383.

Wang Y, Song X, Tian C, et al. Dynamic fractures are an emerging new development geological attribute in water-flooding development of ultra-low permeability reservoirs. Pet Explor Dev. 2015;42(2):222–8.

Wang K, Zhang J, Dai J, et al. Genesis and evolution of reservoir structural fracture in Keshen-2 gas field, Tarim Basin. J Central South Univ (Sci Technol). 2017;48(5):1242–51.

Warren CJ, Miller JMcL. Structural and stratigraphic controls on the origin and tectonic history of a subducted continental margin, Oman. J Struct Geol. 2007;29:541–558.

Wu S. Reservoir characterization & modeling. Beijing: Petroleum Industry Press; 2010. p. 157–61.

Wu J, Wu S, Wei Y, et al. 3D modeling of sedimentary microfacies of Minghuazhen formation of neogene in Gangxi development area in Huanghua Depression. J Palaeogeogr. 2002a;4(4):39–46.

Wu H, Yang F, Ren Y, et al. Gas reservoir evaluated of Xushen-1 Well Area of Xujiaweizi depression in the North of Songliao Basin. Beijing: Petroleun Industry Press; 2002b.

Wu Y, Wu S, Cai Z. Petroleum mineral geology. Beijing: Petroleum Industry Press; 2005.

Wu Y, Chang L, Chen W, et al. Application of fracture characterization and modeling in Dina 2 gas field. Fault-Block Oil Gas Field. 2015a;22(1):78–81.

Wu J, Hu X, Liang Y, et al. Geological formation analysis and evaluation of high radioactive reservoir in beibu gulf basin. Spec Oil Gas Reserv. 2015b; 22(1):79–83.

Xie J. Numerical simulation of Cenozoic tectonic evolution and its genesis in the South China Sea. PhD Dissertation of Graduate School of Chinese Academy of Sciences, Guangzhou Institute of Geochemistry, Chinese Academy of Sciences; 2006.

Xu R, Ao L. Genesis and distribution of Cretaceous salt structures in North Gabon Subbasin. China Petrol Explor. 2016;21(5):70–4.

Xue Y, Lin H, Zhang K, et al. Tectonic characteristics and genetic simulation of Hala'alate Mountain Area. Geotecton Metallog. 2017;41(5):843–52.

Yang K, Yuan Y. Foundation of geology. Wuhan: China University of Geosciences Press; 2009.

Yang Z, Huo L, Zhang Y, et al. Mechanism of non-liner gas seepage flow in water-bearing volcanic gas reservoir. Nat Gas Geosci. 2010;21(3):371–4.

Yang S-J, Duuring P, Kim Y-S. Structural genesis of the Eunsan and Moisan low-sulphidation epithermal Au-Ag deposits, Seongsan district, Southwest Korea. Miner Deposita. 2013;48:467–83.

Yang C, Yang C, Zhang J, et al. Mesozoic tectonic styles and their dynamic mechanisms in the East China Sea Shelf Basin. Mar Sci Bull. 2017;36(4):431–9.

Yuan S, Song X, Ran Q. Fractured reservoir development technology. Beijing: Petroleum Industry Press; 2004.

Zeng L, Ke S, Liu Y. Research method of low permeability oil and gas reservoir fracture. Beijing: Petroleum Industry Press; 2010.
Zhang Y. Oil-controlling characteristics and hydrocarbon accumulation rules of faults in Junggar Basin. Beijing: Petroleum Industry Press; 2010.
Zhang J, Xie J. Development geology. Beijing: Petroleum Industry Press; 2011. p. 60–5.
Zhang X, Chen H, Zhang J, et al. Research technology of structural fracture in metamorphic rock reservoir. Beijing: Petroleum Industry Press; 1999.
Zhang F, Cao G, Li Y, et al. Used of seismic properties analysising technology in researches of volcanic cracks in Xudong Area in the North of Songliao Basin. J Daqing Petrol Inst. 2007a;31(2):12–4.
Zhang F, Song J, Shen Z, et al. A study on fracture orientation and characteristic of remnant paleomagnetization of deep-burial volcanic rocks, north of the Songliao Basin. Chin J Geophys. 2007b;50(4):1167–73.
Zhang H, Zhang S, Wang Y, et al. The history, present situation and future of oil and gas reservoir research. Beijing: Petroleum Industry Press; 2007c.
Zhang Q, Zhai W, Peng J, et al. Research on tectonic causes of numerous ground fissures development mechanism and its unbalance distribution between eastern and western of Weihe Basin. Chin J Geophys. 2012;55(8):2589–97.
Zhang J, Yun J, Zhang Z, et al. The development model of structural fractures in ordovician reservoir in Yubei Area, Tarim Basin. J Geomech. 2014;20(4):413–23.
Zhang W, Wang T, Chang H, et al. Structural characteristics and genetic analysis of Tongcheng fault in Jinhu sag. J China Univ Petrol. 2015;39(5):18–26.
Zhang B, Fu G, Zhang J, et al. Analysis on the different sealing conditions required by mudstone caprock at different fault evolution stages: a case study on the 1st member of Qingshankou Formation in the Sanzhao Depression and 2nd member of Dongying formation in the structure No. 5 of the Nanpu Sag. Oil & Gas Geol. 2017;38(1):22–28.
Zhuang P, Zhizhong C. Fault structure on studying. Beijing: Earthquake Publishing House; 1996.

Chapter 3
Geological Genetic Analysis of Reservoirs

Geological genetic analysis of reservoirs is a crucial part of the geological genetic analysis of oil and gas reservoirs in field development. It includes the determination of the sedimentary environment, origin analysis of different sedimentary facies, origin analysis of different orders of reservoir architectures and single sand-bodies, origin analysis of the barrier/interlayer, origin analysis of the reservoir space, analysis of the main controlling factors of reservoirs, and quantitative classification and evaluation of reservoirs. An accurate understanding of the origin of reservoirs is of great significance for the effective development of oil and gas (Chen et al. 2016b).

3.1 Determination of the Sedimentary Environment of Conglomerate Reservoirs by Grain-Size Analysis

Grain-size analysis is carried out to identify the grain size of clastic rocks and their distribution. Grain-size distribution and sorting of clastic rocks is a measure of the energy of the sedimentary medium, and an effective mark for identifying the natural geographical environment and hydrodynamic conditions in the depositional period. Grain size and its spatial distribution of clastic rocks also influence the reservoir properties; thus, grain-size analysis is also of great significance for reservoir evaluation (Zhu et al. 2008a, b). Many Chinese scholars have devoted great efforts into grain-size analysis (Zhang et al. 2003, 2005; Song et al. 2005; Yuan et al. 2005; Xu et al, 2007; Zhang et al. 2008a, b; Jiang 2009; Zhu et al. 2009; Zhu et al. 2010; Chen et al. 2010a, b; Liu et al. 2010; Zhou et al. 2010; Liu et al. 2011; Pan et al. 2011; Fu et al, 2013; Ge et al. 2018; Yuan et al. 2019). Song et al. (2005) investigated the favorable sedimentary facies belt of the conglomerate reservoir through grain-size analysis according to the heterogeneity and complex modal pore structure of the conglomerate reservoirs in the Karamay Oilfield. Zhang et al. (2005) detailed the application of grain-size analysis in environmental research and stratigraphic division and correlation of the Quaternary in Xiamen, Fujian Province. Zhang

© Petroleum Industry Press 2024

H. Chen, *Geological Genetic Analysis and Application of Oil and Gas Reservoirs*,
https://doi.org/10.1007/978-981-97-7628-3_3

et al. (2008a, b) established the discriminant function of stable lacustrine sediments and typical eolian sediments and put forward its significance as a good quantitative method for analyzing the stable lacustrine sediments and typical eolian sediments of geological history, which is valuable for investigating terrestrial paleoenvironments, drought events, and dust-storm events. Zhu et al. (2010) introduced the application of sediment dynamic analysis in the sedimentary environment interpretation based on a case study of the Upper Cretaceous Honghuatao Formation in the western Jianghan Basin. Chen et al. (2010a, b) analyzed the grain size of the delta sandbody framework in the Upper Triassic Yanchang Formation, North Shaanxi and suggested different grain-size parameters and Sk-Kg in different river patterns. Zhou et al. (2010) introduced the application of clastic grain-size analysis to the genetic study of sand bodies in the Upper Es4 submember of the Xin176 area in the Dongying Sag and indicated that the analysis of the grain texture of sediments-based on grain-size data can effectively determine the transport mode of sediments, judge the hydrodynamic conditions, distinguish the sedimentary environments, and analyze the genetic mechanism of sediments. Pan et al. (2011) analyzed the characteristics of grain-size parameter distribution, grain-size distribution curves, and log-probability cumulative curves using the grain-size data of the Core SE2 in the southern bank of the Qiantang River and interpreted the environmental evolution in this area since the Late Quaternary according to the lithology, sedimentary structures, and foraminifera distribution. Ge et al. (2018) explored the application of grain-size analysis in the evaluation of the sedimentary environment, with a case study on the tight sandstone reservoir of the Taiyuan Formation in the Linxing area, the Ordos Basin. Yuan et al. (2019) reviewed the history of grain-size analysis, summarized the commonly used types of grain sizes, introduced the grain-size standards and five grain-size analysis methods usually used in sedimentology presently, and described a detailed classification of the sedimentation method, field interference analysis method, and image method. Previous studies adopted grain-size analysis mainly for sedimentary environment analysis and stratigraphic division and correlation. They focused on fluvial facies and delta facies, but rarely on alluvial fans. In this section, the sedimentary environment of the alluvial-fan reservoirs of the Lower Karamay Formation in an area at the northwestern margin of the Junggar Basin is discriminated by grain-size analysis, and the results are verified using the data of sedimentary structures and logging facies.

3.1.1 Foundations for Discriminating the Sedimentary Environment of Conglomerate Reservoirs by Grain-Size Analysis

The Karamay Oilfield is located at the northwestern margin of the Junggar Basin and is adjacent to the Zaire Mountain in the west. Distributing as a NE-SW band, approximately 50 km long and 10 km wide, the oilfield is a monocline structure, descending in a steplike form from NW to SE (Fig. 3.1). The study area is separated

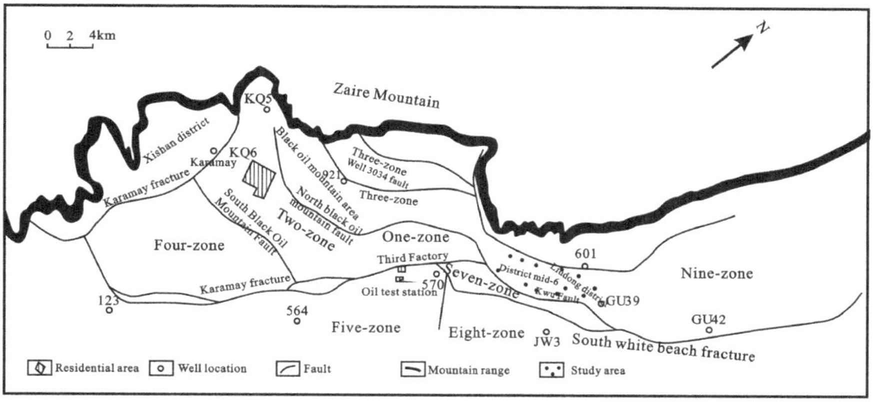

Fig. 3.1 Structural map of Karamay oilfield in the Junggar Basin (After Zheng et al. 2010b)

into nine zones and several fault blocks by faults. It is bounded by Zone 7 and the Karamay-Wuerhe fault to the southwest and adjacent to Zone 9 in the east and is divided into two blocks by the Baijiantan fault. From bottom to top, the study area develops the Carboniferous, Triassic, Jurassic, and Cretaceous. The Triassic includes the Baikouquan Formation, Lower Karamay Formation, Upper Karamay Formation, and Baijiantan Formation, where the Lower and Upper Karamay Formations are the main oil-bearing series. The Lower Karamay Formation, of which nine layers (S63, S71, S721, S722, S723, S731, S732, S733, and S74) are the targets in this study, has a burial depth of 350–850 m and formation thickness of 50–70 m. There are 1085 injectors and producers in the study area, with an average well spacing of approximately 110 m (Zheng et al. 2010a, b). The Triassic is a very thick (300–2500 m) set of grayish-green–brownish-red conglomerates. The Lower Triassic, only observed in the eastern part of the oilfield, is almost completely occupied by conglomerates and psephitic sandstones, with a thickness of 130–200 m. The Middle Triassic is widespread, consisting of a fining-upward sedimentary succession from conglomerate–sandstone, conglomerate interbedded with mudstone, to siltstone–mudstone in the lower part, and a 50–450-m thick set of glutenite alternating with mudstone in the upper part (Jiang 2003).

3.1.2 Grain-Size of the Conglomerate Reservoir

3.1.2.1 Core Observation of Grain Size

The lithologic interpretation results through core observation and grain-size analysis are used to define the lithologic properties and provide support for the determination of sedimentary facies. Then, sedimentary microfacies are analyzed according to the correspondence between microfacies and lithology. The results of core observation and grain-size analysis (or sieving analysis) of key wells illustrate that the Lower Karamay Formation is preliminarily composed of medium sandstone, sandy conglomerate, and fine conglomerate, with fine sandstone and mudstone locally (Fig. 3.2). Their grains are coarse (generally > 2 mm) and poorly rounded, mostly appearing subrounded to subangular. The sediments usually occur as a mixture of sands, gravels, and muds, with poor sorting. The Lower Karamay Formation develops fining-upward sediments and are roughly divided into three intervals—S74, S733–S723, and S722–S63.

(1) S74

This interval is dominated by glutenite, with bands of medium conglomerate and fine conglomerate. It corresponds to channel flow and laminar-flow belts and includes the microfacies of channel-flow gravel bodies, channel-shoal sand/gravel bodies, and laminar-flow gravel bodies.

(2) S733–S723

Fig. 3.2 Grain size characteristics of reservoir core observation of Kexia Formation in northwestern margin of Junggar Basin. **a** J1 Well, dark brown conglomerate, 398.61–398.80 m; **b** J7 well, light brown sandy conglomerate with medium gravel, 421.44–421.60 m; **c** J1 well, normal-grained, light brown sandy conglomerate with medium gravel, 402.51–402.80 m; **d** J7 well, light grayish brown fine conglomerate, 434.64–434.80 m; **e** J7 well, light gray mudstone—brown fine sandstone, stratum mutation contact, 417.04–417.15 m; **f** J1 well, light gray green silty mudstone, 397.66–397.79 m

This interval is dominated by sandstone, with bands of glutenite and fine conglomerate. It corresponds to a braided flow belt and includes the microfacies of the braided channel and braided sand/gravel bar.

(3) S722–S63

This interval is dominated by mudstone, with bands of sandstone and glutenite. It also roughly corresponds to a braided flow belt and includes the microfacies of the braided channel and braided sand/gravel bar.

Based on the correspondence between microfacies and lithology described above and the sedimentary structures and logging facies, the sedimentary environment can be subdivided by microfacies to realize fine description.

3.1.2.2 Grain-Size Histogram

A grain-size histogram can intuitively show the variation of grain size and percentage of clasts of different grain sizes. The grain size of conglomerate reservoirs in the Lower Karamay Formation has a wide distribution range, and the mass percentage of the peak grain sizes is not high (Fig. 3.3). This indicates that the reservoir rocks are generally poorly sorted, which agrees with the basic features of the alluvial-fan conglomerate reservoir, and provides some support to the determination of alluvial-fan sediment types.

3.1.3 Application of Grain-Size Analysis in the Discrimination of the Sedimentary Environment

Grain size of sedimentary rocks is controlled by the transport medium, transport mode, sedimentary environment, and other factors, which in turn are bound to be reflected by the grain size. This is exactly evidence for defining the sedimentary environment by grain-size analysis (Zhu 2008a, b).

3.1.3.1 Grain-Size Log-Probability Cumulative Curve

The use of grain-size log-probability cumulative curves to establish the typical model of a sedimentary environment was proposed by Visher (Galloway et al. 1983; Zhu 2008a, b). Here, grain-size log-probability cumulative curves are plotted for the typical rock types in the conglomerate reservoirs of the Lower Karamay Formation at the northwestern margin of the Junggar Basin. These curves illustrate that the rocks mainly consist of coarse-grained components transported by rolling and saltation, but rare fine-grained components transported by suspension, because the conglomerate reservoirs are generally coarse-grained. This fully reveals the characteristic

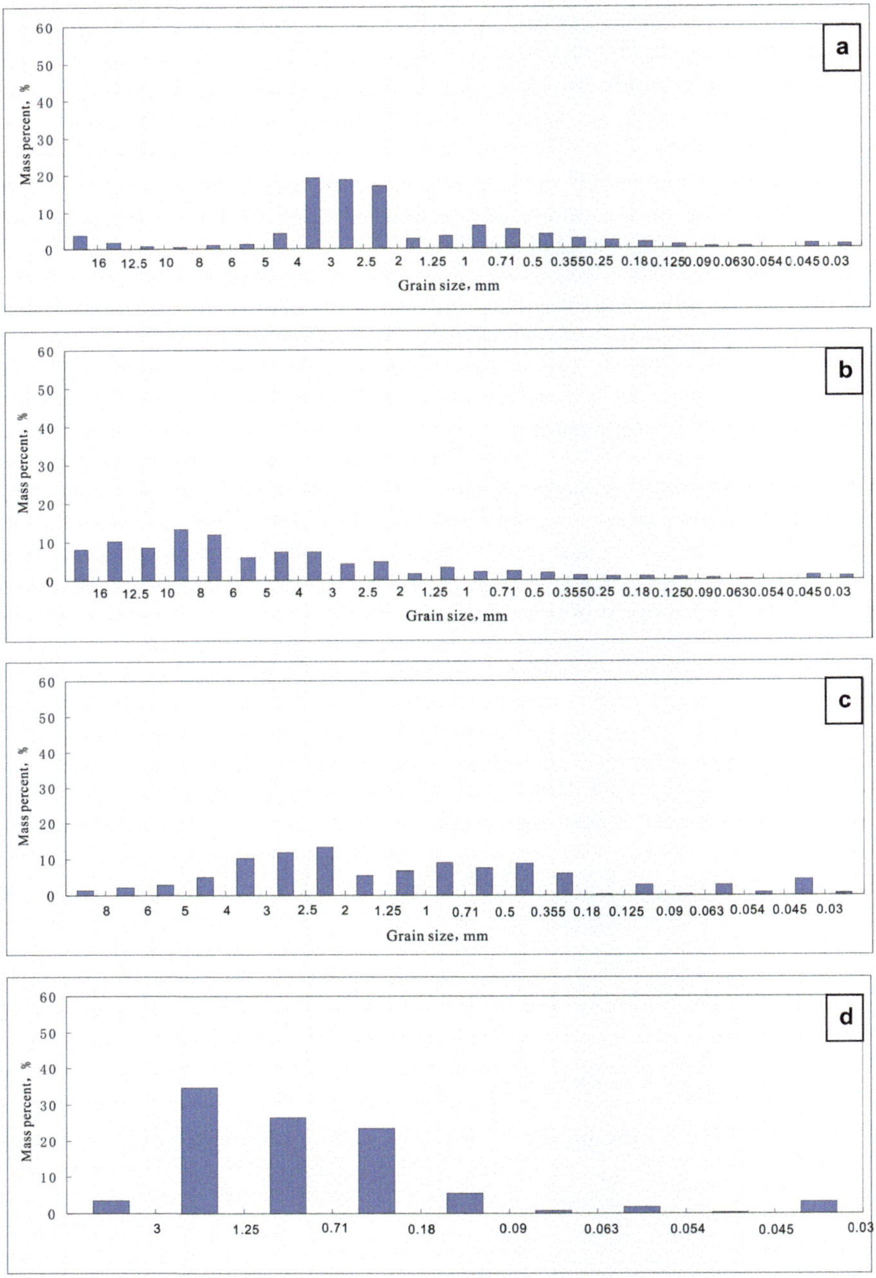

Fig. 3.3 Granularity components histogram of different kinds of rocks of lower-Karamay Formation in Karamay oilfield in the Junggar Basin. **a** J3 well, sandy fine conglomerate grain size composition histogram, 406.20 m; **b** J3 well, 422.40 m; histogram of grain size composition of conglomerate in sand; **c** J55 well, glutenite grain size composition histogram, 396.24 m; **d** J56, grain size composition histogram of gravelly unequal-grained sandstone, 404.44 m

dominance of coarse-grained sediments in the alluvial-fan conglomerate reservoirs and provides a basis for sedimentary environment discrimination. As poorly sorted, the curves all show a relatively small slope in Fig. 3.4. Sandy fine conglomerate and glutenite are mainly developed in the braided flow belt of the alluvial fan. Influenced by the current scour, a saltation population is divided into two subpopulations, showing as two intersecting line segments that slightly differ in median size and sorting.

3.1.3.2 C–M Diagram

A *C*–*M* diagram is plotted with *C* and *M* values of a sample. *C* is the grain diameter corresponding to 1% on the log-probability cumulative curve and equivalent to the diameter of the coarsest grain in the sample; it represents the maximum hydrodynamic energy allowable for sediment transportation. *M* is the median grain diameter, representing the average hydrodynamic energy. A *C*–*M* diagram of the Lower Karamay Formation alluvial-fan reservoirs plotted with data for 133 samples from four wells in S7 (Fig. 3.5). It is seen that the Lower Karamay Formation is mainly traction-current deposits, where only a short line segment is parallel to the $C = M$ baseline. The samples are dominantly coarse-grained sandy conglomerates and glutenites, showing as O–P and N–O segments, and are sediments of channel flow, braided channel, and braided sand/gravel bar of the alluvial fan transported by traction.

3.1.3.3 Scatter Diagram of the Structural Parameter

Mason and Folk believed that different sedimentary environments possess different grain-size parameters and proved that the scatter diagram of skewness and kurtosis is a very effective tool in distinguishing a beach, a coastal dune, and a wind flat. The scatter diagram developed by Friedman using the moment-based standard deviation and skewness of representative sand samples collected worldwide can clearly distinguish river sand, sea-beach sand, and lake-beach sand (Zhu et al. 2008a, b). Figure 3.6 illustrates a scatter diagram of the skewness and standard deviation of 180 samples collected from four wells in the Lower Karamay Formation in the northwestern Junggar Basin. It is seen that the alluvial-fan conglomerate reservoirs have positive skewness and low degree of sorting, which reveals a sedimentary environment with the dominance of traction current formed by temporary torrent on the alluvial fan, as deposits of the main channel in the upper fan or braided channel radiating toward the lower fan. Gradually transiting from a wet fan to a dry fan from bottom to top, the alluvial fan of the target formation is a fining-upward regressive succession. This feature of traction-current sediment is especially remarkable in the middle-upper part of the target formation, that is, from S723 to S63. The traction-current sediments in the target formation reflect the dominance of rolling and saltation populations on the log-probability grain-size distribution curve, different from general fluvial

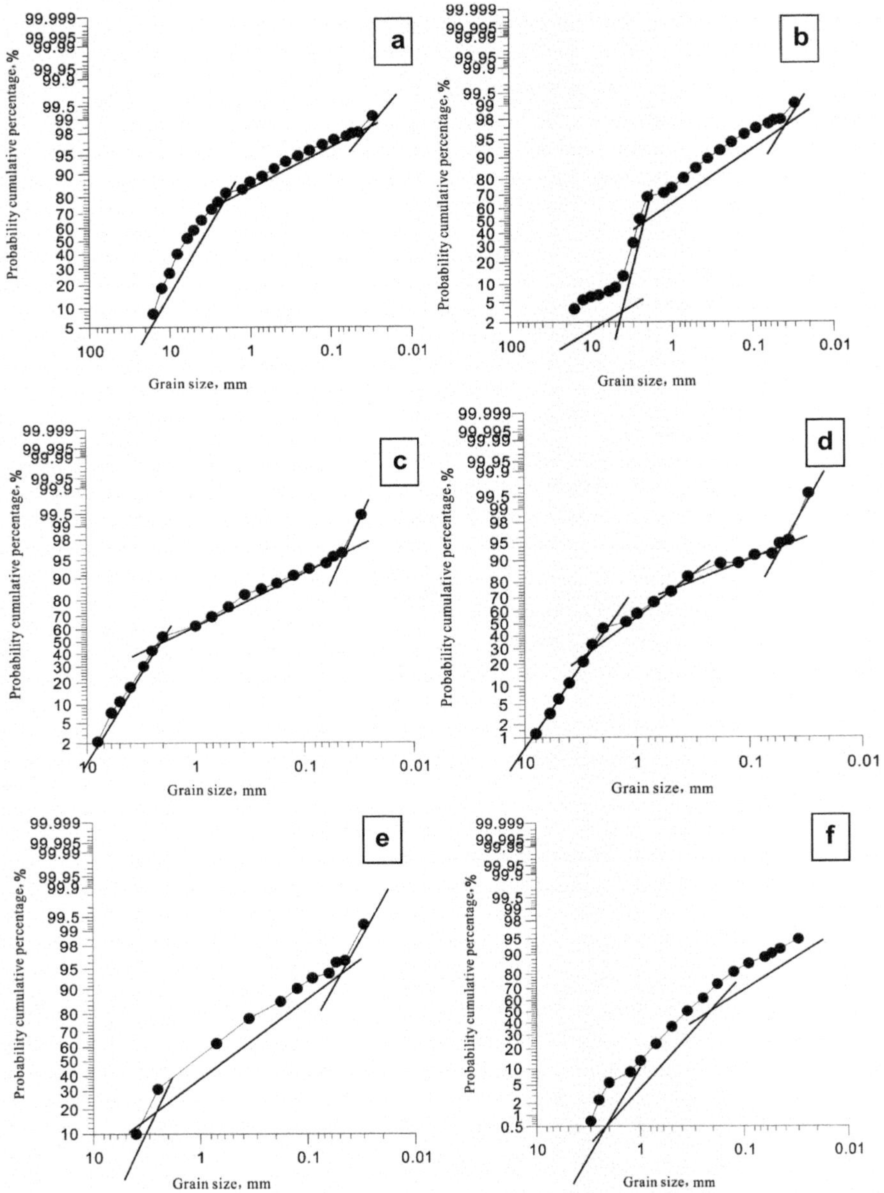

Fig. 3.4 Probability map characteristics of mainly kinds of rocks of lower-Karamay Formation in Karamay oilfield in the Junggar Basin. **a** J3 well, sand-bearing conglomerate grain size probability cumulative curve, 422.40 m; **b** J3 well, grain size probability cumulative curve of sandy fine conglomerate, 406.20 m; **c** J55 well, glutenite grain size probability cumulative curve, 396.24 m; **d** J56 well, cumulative probability curve of grain size of unequal grain sandstone, 387.15 m

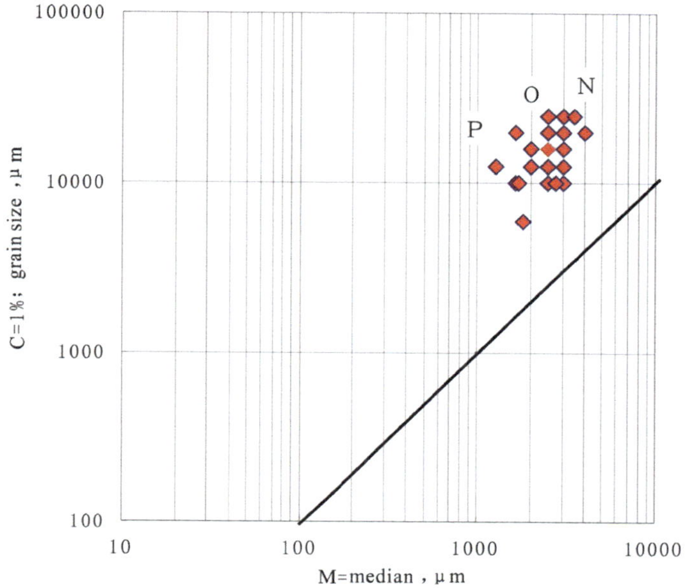

Fig. 3.5 C–M map characteristics of alluvial fan reservoir of lower-Karamay Formation in Karamay oilfield in the Junggar Basin (Well-J581, sandy-conglomeratic)

traction-current sediments, which generally demonstrate a few suspension population, primarily due to the dominance of coarse debris supply from the proximal source. The characteristics of the sedimentary environment described above clearly influence reservoir properties. As the sedimentary environment changes vertically, from the inner zone of the upper fan to the outer zone of the upper fan, then to the middle fan, and finally to the lower fan, the grain size of sediments declines gradually. Consequently, the reservoir properties become worse from bottom to top. The inner and outer zones of the upper fan are the major pay zones currently, and the middle fan is a replacement for further development.

According to the grain-size analysis above, the Lower Karamay Formation in the northwestern Junggar Basin is defined as typical alluvial-fan conglomerate reservoirs, which are coarse-grained and poorly sorted, mainly consisting of medium sandstone, sandy conglomerate, and fine conglomerate. Note that grain-size analysis can provide information on the sedimentary environment, especially on hydrodynamics, but its results may not necessarily be ideal due to the complex origins of the sedimentary environment. Overall, different sedimentary environments have different hydrodynamic conditions, but similar hydrodynamic conditions may appear in the subenvironments of different environments. Additionally, differences in sediment supply, structural conditions, and other factors always result in great complexity.

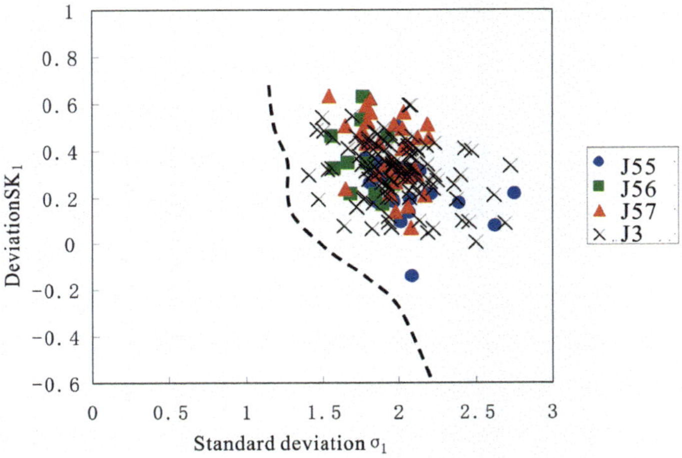

Fig. 3.6 Scatter diagram of skewness and standard structure parameters of alluvial fan reservoir of lower-Karamay Formation in Karamay oilfield in the Junggar Basin (180 samples from 4 wells)

Therefore, the only way to draw objective and true conclusions is to combine grain-size analysis with other relevant data and methods for cross-verification. Thus, sedimentary structure analysis, logging facies analysis, and other means of analyzing the sedimentary environment are used in this study as an effective supplement to grain-size analysis and an approach to verify its results. By analysis of sedimentary structures in core samples, graded bedding and quasi-parallel bedding are identified in laminar-flow gravels and cross bedding and scour surface are identified in braided channels (Fig. 3.7). In logging facies analysis, channel-flow gravel bodies and laminar-flow gravel bodies show funnel-shaped or inverted trapezoidal-logging curves, braided channels exhibit the characteristic bell-shaped or box-shaped logging curves of the normal graded structure, and cross-flow fine grains generally have low and flat electrical-logging curves (Table 3.1).

3.2 Sedimentological Analysis of the Geological Origin of Reservoirs

As a representative type of transitional sedimentary facies, fan delta is common in many sedimentary basins and rich in oil and gas; thus, it has always been one of the main research targets (Galloway et al. 1983; Mei 1994). By definition, fan delta is the alluvial-fan prograding from the adjacent highlands into the sea, lake, and other stable water bodies (Jiang et al. 2003). Many scholars have worked extensively on fan delta (Chen et al. 2006; Wang and Zhong 2010; Lin et al. 2013; Yu et al. 2014; Tang et al. 2014; Yang et al. 2015; Zhu et al. 2015; Zhang et al. 2015a, b; Song

Fig. 3.7 Core bedding characteristics of Kexia Formation in northwestern margin of Junggar Basin. **a** J1 well, grain sequence bedding, gray black coarse conglomerate, sheet flow gravel body, 419.24–419.40 m; **b** J7 well, parallel bedding, brown pebbly coarse sandstone, sheet flow gravel body, 426.1–426.4 m; **c** J1 well, scour surface, brown sandy fine conglomerate—light gray green pebbly argillaceous sandstone, 405.81–406.02 m; **d** J1 well, plate cross bedding, light gray green medium fine sandstone—light gray green argillaceous siltstone 406.51–406.72 m

et al. 2015). Chen et al. (2006) established a prototype framework model of fan-delta fronts using the data of closely spaced wells in the Shuanghe Oilfield. Furthermore, Wang and Zhong (2010) provided a detailed description of the sedimentary features of fan delta through a field survey on fan-delta outcrops of the Middle Jurassic Dameigou Formation in the south margin of the Altun Mountain, western Qaidam Basin; the Lower Cretaceous Miaogou Group in Hongliugou, Chaoshui Basin; and the Jurassic in Sangyuanyingzi, Luanping Basin. Lin et al. (2013) anatomized the architectures of reservoirs in fan-delta fronts, analyzed the characteristics of architectural elements, and established a 3D reservoir architecture model, with the Dujiatai reservoir in the Shu2-6-6 block, Liaohe Oilfield, as an example. Yu et al. (2014) studied the lithofacies and origin models of the fan-delta conglomerate of the Baikouquan Formation in the Mahu Sag. Furthermore, Tang et al. (2014) analyzed the characteristics and distribution of the fan-delta group of the Baikouquan Formation in the Mahu Sag, classified the target formation as shallow-water fan delta deposited under the background of gentle slope, and identified the development of gravity-flow and traction-current lithofacies. Additionally, Yang et al. (2015) analyzed the origin of high-quality reservoirs of the fan-delta front in the lower part of the fourth member

Table 3.1 Classification characteristics of configuration unit of alluvial fan conglomerate reservoir

Subfacies	Four-order architectures	Lithology	SP	RT
Inner zone of the upper fan	Channel-flow gravel bodies	Coarse conglomerate, medium conglomerate, gravel-bearing glutenite	Large, funnel-shaped, and inverted trapezoidal	High, funnel-shaped, and inverted trapezoidal
	Channel-shoal sand/gravel bodies	Medium conglomerate, glutenite	Large–medium, funnel-shaped, and inverted trapezoidal	Medium–high, funnel-shaped, and inverted trapezoidal
	Sheetflood inner sand bodies	Gravel-bearing sandstone, coarse sandstone	Medium–small, funnel-shaped, and inverted trapezoidal	Low–medium, funnel-shaped, and inverted trapezoidal
	Sheetflood inner fine grains	Siltstone, argillaceous siltstone, gravel-bearing mudstone	Low, straight	Low, straight
Outer zone of the upper fan	Laminar-flow gravel bodies	Medium conglomerate, fine conglomerate containing medium conglomerate, argilliferous glutenite	Large, dominantly funnel-shaped, and inverted trapezoidal	High, dominantly funnel-shaped, and inverted trapezoidal
	Sheetflood outer sand bodies	Gravel-bearing sandstone, medium–fine sandstone	Medium–small, funnel-shaped, and inverted trapezoidal	Low–medium, funnel-shaped, and inverted trapezoidal
	Sheetflood outer fine grains	Siltstone, argillaceous siltstone, gravel-bearing mudstone	Low, straight	Low, straight
Middle fan	Braided channel	Sandyconglomerate, gravel-bearing sandstone, coarse sandstone, medium sandstone	Medium–large, bell-shaped, and box-shaped	Medium–high, bell-shaped, and box-shaped
	Braided sand/gravel bar	Fine conglomerate, sandy conglomerate, gravel-bearing sandstone	Large–medium, funnel-shaped, and inverted trapezoidal	Medium–high, funnel-shaped, and inverted trapezoidal
	Cross-flow sand bodies	Gravel-bearing sandstone, fine sandstone	Low–medium, bell-shaped, and box-shaped	Medium–high, finger-shaped

(continued)

Table 3.1 (continued)

Subfacies	Four-order architectures	Lithology	SP	RT
	Cross-flow fine grains	Gravel-bearing siltstone and gravel-bearing mudstone	Low, straight	Low, straight, or jagged
Lower fan	Runoff channel	Fine sandstone, siltstone	Low–medium, bell-shaped, and box-shaped	Medium–high, finger-shaped
	Interchannel fine grains	Silty mudstone, mudstone	Low, straight	Low, straight, or jagged

of the Shahejie Formation, Bonan Subsag. Zhu et al. (2015) discussed the diagenesis and origin of the favorable reservoirs of the fan delta in the third member of the Shahejie Formation of the Eocene in the Zhanhua Sag of the Jiyang Depression and indicated that diagenesis improves the reservoir properties and that the depth interval where secondary pores exist is favorable for high-quality reservoir formation and hydrocarbon accumulation. Zhang et al. (2015a, b) reviewed the literature on fan-delta sedimentology and presented the advances in research from four aspects, namely, classification of fan-delta sedimentary systems, lithofacies and sedimentary sequence, depositional models, and research methodology. Song et al. (2015) identified the architecture boundaries of the fan-delta front reservoirs in the Dujiatai reservoir in the Jin-99 block of the Huanxiling Oilfield, Liaohe Basin, providing a basis for a delicate study of microfacies. Previous studies focused on establishing a sedimentary facies model, the classification of sedimentary microfacies, and genetic analysis of reservoirs based on logging, outcrop, and core data. However, few studies have been conducted on fan-delta reservoirs sedimentologically to deeply understand the spatial development of reservoirs. In this section, the sedimentary microfacies in reservoirs are clarified by a comprehensive analysis of sedimentary features of the fan-delta front through core observation, thin-section analysis, and tests of samples taken from seven wells, and using the logging data of 400 wells and seismic data of the study area to provide a geological basis for the conversion from steam huff and puff to steam flooding for heavy oil recovery.

3.2.1 Foundations for the Sedimentological Analysis of the Geological Origin of Reservoirs

The study area is structurally located at the southern end of the western slope of the Western Sag of the Liaohe Basin and appears as a SE-trending monocline (Fig. 3.8) (Li et al. 2010a, b; Chen et al. 2015a, b). The Yulou reservoir in the first member of the Shahejie Formation (Sha 1 Member) of the Paleogene, the target in this study, exhibits a SE-trending monoclinal structure, with dip angles between 2° and 10°. It is a clastic deposit of fan-delta front and consists mainly of thick inequigranular sandstone and medium–fine sandstone. Reservoirs in the study area have high porosity (avg. 31.25%) and high permeability (avg. 1829.3 mD). The materials were supplied from the northwest and north. Sedimentary microfacies in the Yulou reservoir include an underwater distributary channel, river-mouth bar, delta-front sheet sand, underwater distributary interchannel sand, and underwater distributary interchannel mud, with an underwater distributary channel, river-mouth bar, and underwater distributary interchannel sand in dominance. The study area has been developed roughly in four stages by recovery mode, that is, pilot test of steam huff-and-puff, full implementation of steam huff-and-puff, comprehensive treatment by reducing well spacing, and pilot test of steam flooding in the middle to late stage of steam huff-and-puff. The Yulou reservoir is currently producing from two series of layers vertically—in the northeastern part, steam flooding has been implemented with good results; in the rest parts, steam huff-and-puff is conducted at the middle to late stage and yields worse and worse production performance, suggesting the necessity of converting the development mode. The Yulou reservoir is a moderately thick layered heavy oil reservoir with edge/bottom water. The viscosity of yI oil degassed on the surface at 50°C is 11,577 mPa s. The frequent swing of underwater distributary channels led to vertical superimposition and horizontal alternation of different phases of the channel and interchannel sand bodies, thus allowing the reservoir to be highly heterogeneous to restrict the expansion of steam flooding. Here, the reservoir will be characterized through a systematic analysis of sedimentology, aiming to provide a reference for effective reservoir development. The Yulou reservoir in the study area is divided into 29 sublayers ($yI_1{}^{1a}$, $yI_1{}^{1b}$, $yI_1{}^{2a}$, $yI_1{}^{2b}$, $yI_1{}^{2c}$, $yI_2{}^{3a}$, $yI_2{}^{3b}$, $yI_2{}^{3c}$, $yI_2{}^{4a}$, $yI_2{}^{4b}$, $yI_2{}^{4c}$, $yI_3{}^{5a}$, $yI_3{}^{5b}$, $yI_3{}^{5c}$, $yI_3{}^{6a}$, $yI_3{}^{6b}$, $yI_3{}^{6c}$, $yII_1{}^{1a}$, $yII_1{}^{1b}$, $yII_1{}^{2a}$, $yII_1{}^{2b}$, $yII_2{}^{3a}$, $yII_2{}^{3b}$, $yII_2{}^{4a}$, $yII_2{}^{4b}$, $yII_3{}^{5a}$, $yII_3{}^{5b}$, $yII_3{}^{6a}$, and $yII_3{}^{6b}$), corresponding to 29 short-term base-level cycles (Fig. 3.9; Chen et al. 2014). The establishment of a layer isochronal stratigraphic framework makes the subsequent sedimentary microfacies analysis possible at the level of the layer with greatly enhanced accuracy, meeting the requirements of practical development and production of the oilfield.

3.2.2 *Sedimentary Features of Fan-Delta Front*

The sedimentary facies of the Yulou reservoir in the study area should be determined before clarifying its sedimentary features. For this purpose, a comprehensive analysis is conducted considering lithology, bedding, texture of sediments, mudstone color, paleontology, and the grain size of sediments.

3.2.2.1 Lithology

Based on the detailed observation and description of cores with a total length of 668 m collected from seven wells (Fig. 3.10a–d), together with the grain-size data

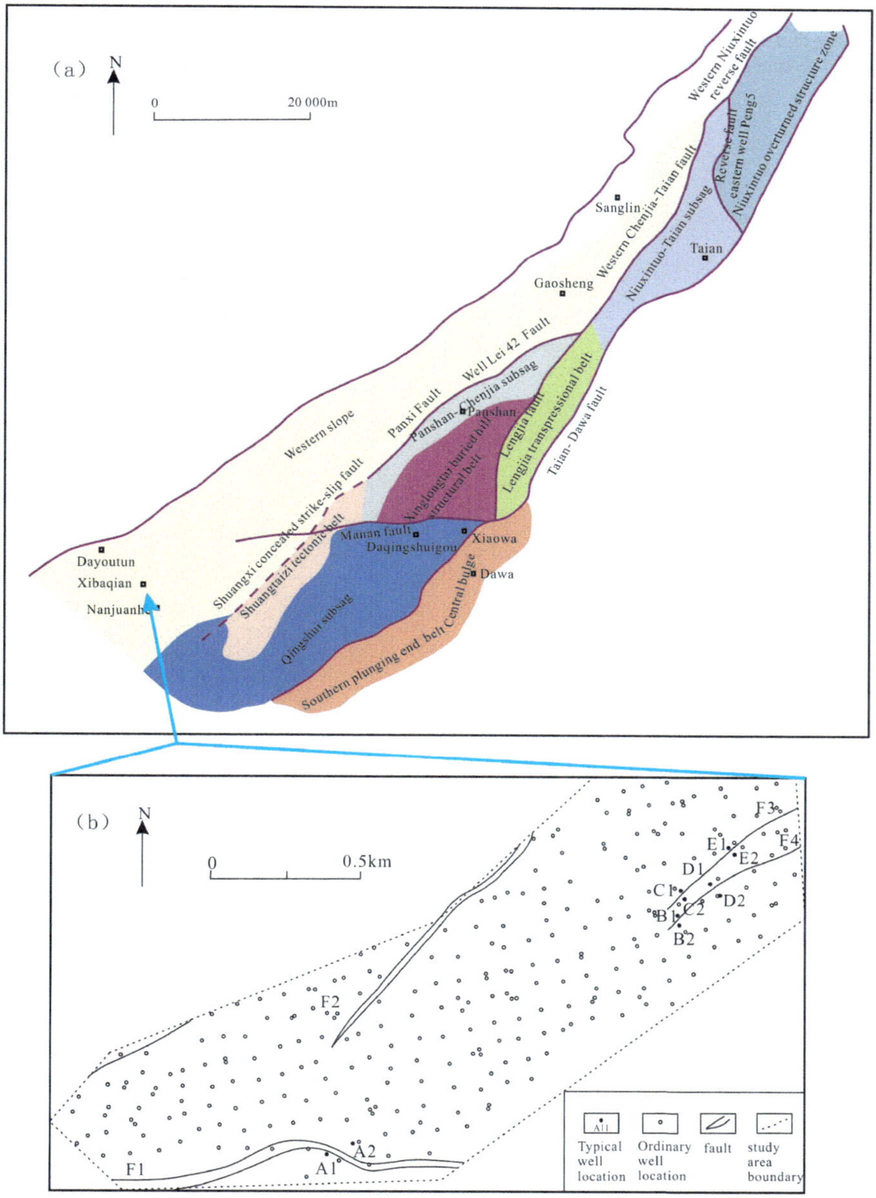

Fig. 3.8 Structural division and research area location map of the western depression of Liaohe Basin (Li Minggang et al. 2010b; Chen et al. 2015b). **a** The structural zoning map of the western depression; **b** Location map of study area

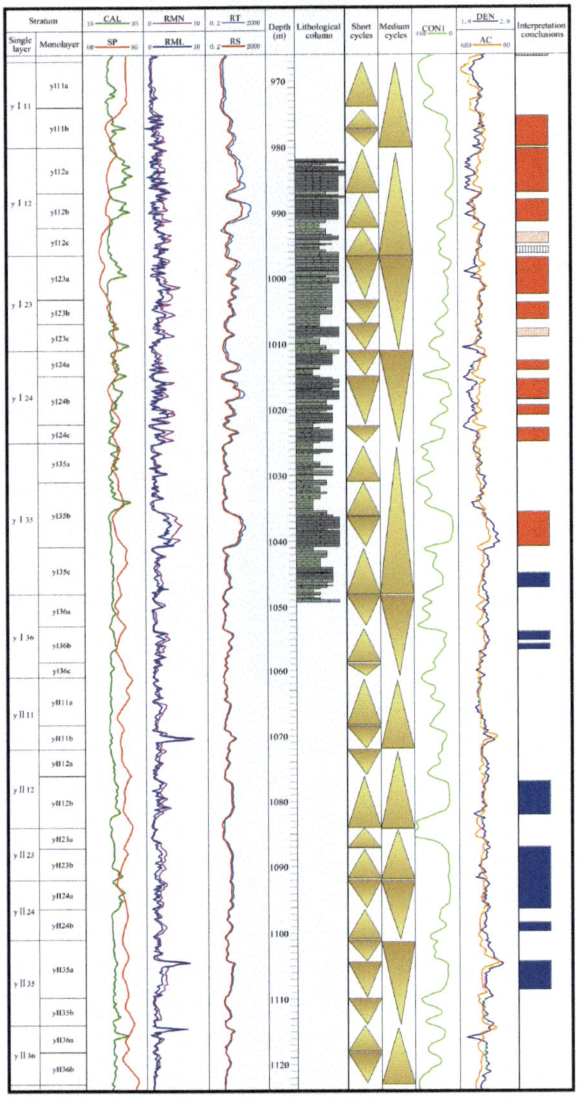

Fig. 3.9 Finely divided and contrasted results of aimed stratum in study area

of 233 samples in the study area, it is indicated that the Yulou reservoir contains various types of rocks, including fine conglomerate, glutenite, coarse sandstone, medium sandstone, fine sandstone, siltstone, argillaceous siltstone, silty mudstone, and mudstone, which are dominantly gray, dark gray, and grayish green. Moreover, the Yulou reservoir presents a large proportion of mixed sands and gravels, with high shale content. Generally, medium and fine sandstones are dominant, consisting of coarse-grained sediments.

Fig. 3.10 Core photo characteristics of sedimentary facies marker of Yulou oil layer in a certain area of western sag of Liaohe Basin. **a** J22-10,944.73–944.78 m, grayish brown fine conglomerate; **b** J2,945.39–945.53 m, grayish-brown fine sandstone; **c** J2,948–948.1 m, siltstone; **d** 10–22,1040.8–1041 m, mudstone; **e** A91,1037.18–1037.28 m, gray-black siltstone, trough cross-bedding; **f** A10 well, 931.44–931.64 m, argillaceous siltstone, tabular cross-bedding; **g** grey-black mudstone, horizontal bedding, 23–261, depth: 999.32–999.36 m; **h** J2, argillaceous siltstone, cephalopod fossils (snails), depth: 979–979.15 m

3.2.2.2 Bedding

Controlled by sediment supply, sedimentary environment, and hydrodynamic conditions, the Yulou reservoir develops a great variety of beddings, including trough cross bedding, tabular cross bedding, wavy bedding, flaser bedding, parallel bedding, and horizontal bedding. The diversity of bedding indicates the continuous change in hydrodynamic conditions during the deposition (Fig. 3.10e, f). These beddings further enhanced the reservoir heterogeneity, which is visually reflected in cores (Fig. 3.10a, e). Carbonaceous mudstone at the bedding interface has a significant influence on the spatial connectivity of the reservoir.

3.2.2.3 Texture of Sediments

As close to the origin, sediments are transported in a short distance, with a mixture of sands and gravels. They are generally coarse-grained, with a high shale content, poor sorting and roundness, and low compositional maturity of minerals (Fig. 3.11).

Fig. 3.11 Sediment composition and structural characteristics of Yulou oil layer in a certain area of western sag of Liaohe Basin. **a** J22-10,1029.5–1029.6 m, gray with medium gravel fine conglomerate; **b** J22-10,1029.9–1030 m, gray with medium gravel fine conglomerate; **c** A261,1027–1027.1 m, gray fine conglomerate; **d** J2,947.05–947.15 m, grayish brown glutenite

3.2.2.4 Mudstone Color

Mudstone in the Yulou reservoir has variable colors, including dark gray, gray, and grayish green. It reveals a reducing environment and the dominance of underwater sediments (Fig. 3.10d, g).

3.2.2.5 Paleontology

Much cephalopoda fossils (snails) are found in mudstones and siltstones (Fig. 3.10h), reflecting the sedimentary features of the fan-delta front.

3.2.2.6 Grain Size of Sediments

The log-probability grain-size distribution curves plotted for the Yulou reservoir show two-segment and three-segment patterns (Fig. 3.12), indicating sediments transported by traction, saltation, and suspension before deposition. This reflects the sedimentary environment of the underwater distributary channel of the fan delta with composite origin.

As described above and in accordance with previous studies (Yu et al. 1999; Sun et al. 2001; Bao et al. 2009), the Yulou reservoir in the study area is classified as fan-delta front sediments, where medium–fine sandstones are dominant. The mudstone color reflects a reducing environment, and the reservoirs are dominated

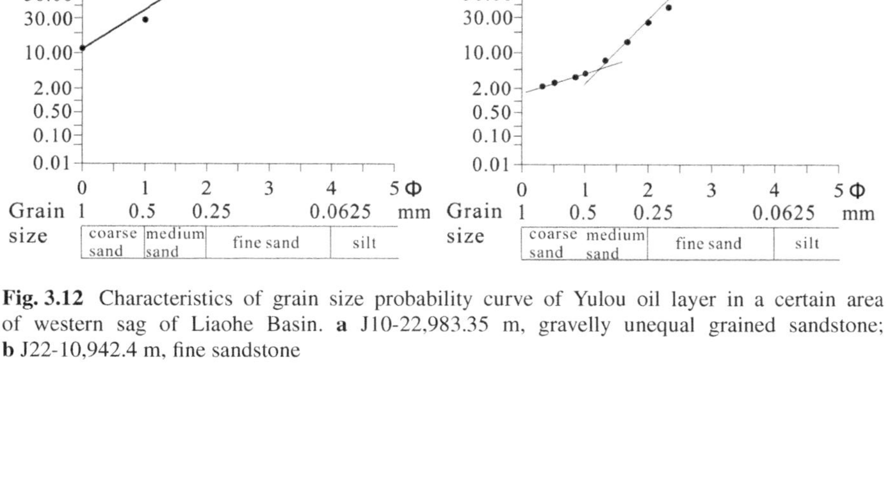

Fig. 3.12 Characteristics of grain size probability curve of Yulou oil layer in a certain area of western sag of Liaohe Basin. **a** J10-22,983.35 m, gravelly unequal grained sandstone; **b** J22-10,942.4 m, fine sandstone

by underwater distributary channel sediments followed by underwater distributary interchannel sands.

3.2.3 Classification of Sedimentary Microfacies

According to the characteristics of lithology, sedimentary origin, sedimentary structure, and logging facies identified depending on the core observation results, logging data of noncoring wells, and testing of well data, the fan-delta front subfacies in the Yulou reservoir is divided into an underwater distributary channel, a river-mouth bar, a delta-front sheet sand, an underwater distributary interchannel sand, and underwater distributary interchannel mud microfacies (Table 3.2).

3.2.3.1 Underwater Distributary Channel

An underwater distributary channel is fining upward, lithologically varying from gray fine conglomerate to dark brown, grayish brown, and gray gravel-bearing coarse–medium sandstone, and then to light brown, grayish green, and gray white fine sandstone and siltstone. It is mainly coarse-grained, moderately sorted, and moderately rounded. Its sedimentary structures reflect multiple types of bedding, including cross bedding and parallel bedding, with a scour surface and channel-lag deposits observed at the bottom (Fig. 3.13). Gravels, which are mainly muddy gravels and rarely sandy gravels, are arranged along beddings. The RT (resistivity) curve is box- or bell-shaped, with abrupt contact at the top and bottom, with RT > 7 W m (Fig. 3.14). This microfacies is the most dominant in the Yulou reservoir.

3.2.3.2 River-Mouth bar

The river-mouth bar is coarsening upward, lithologically consisting of light brown and gray siltstone, fine sandstone, and medium sandstone. It is well sorted and rounded. Its sedimentary structures are medium- to small-scale cross bedding, parallel bedding, wavy cross bedding, and lenticular bedding, as well as deformation bedding and bioturbation structures in silty mudstones. The SP (spontaneous potential) curve is funnel-shaped, and the RT curve is serrated funnel-shaped or high-amplitude polymodal finger-shaped, with RT > 5 W·m. Frequent diversions of underwater distributary channels make the river-mouth bar impossible to preserve in the depositional process; thus, this microfacies is rarely observed in the Yulou reservoir (Fig. 3.14).

Table 3.2 Microfacies in the Yulou reservoir in an area of the Liaohe Basin's West Sag

Subfacies	Microfacies	Rock types	Sedimentary origin	Sedimentary structure	Logging facies	Logging curve shape SP	Logging curve shape RT
Fan-delta front	Underwater distributary channel	Medium conglomerate, fine conglomerate, gravel-bearing sandstone, coarse sandstone, medium sandstone, fine sandstone	Vertical accretion, down-flow accretion, lateral accretion, and aggradation	Parallel bedding, tabular cross bedding, trough cross bedding, graded bedding, scour surface	Box-shaped, bell-shaped		
	River-mouth bar	Gravel-bearing sandstone, coarse sandstone, medium sandstone, fine sandstone	Vertical accretion, down-flow accretion, lateral accretion	Parallel bedding, tabular cross bedding, trough cross bedding	Funnel-shaped		
	Delta-front sheet sand	Medium sandstone, fine sandstone, siltstone	Selective deposition	Ripple cross bedding, flaser bedding, wavy bedding, lenticular bedding, parallel bedding	Finger- or pinnacle-shaped		
	Underwater distributary interchannel sand	Fine sandstone, siltstone, argillaceous siltstone	Lateral accretion, flooding deposition	Ripple cross bedding, flaser bedding, wavy bedding, lenticular bedding	Flat or slightly fluctuated, finger- or pinnacle-shaped		
	Underwater distributary interchannel mud	Silty mudstone, mudstone	Flooding deposition	Horizontal bedding	Straight		

Fig. 3.13 Core characteristics of sedimentary microfacies of Yulou oil layer in a certain area of western sag of Liaohe Basin. **a** grey fine conglomerate, well A261, depth: 1027–1027.1 m; **b** grey-black fine sandstone, well A91, 1037.18–1037.28 m; **c** conglomerate fine sandstone, J2 well, retention deposition, depth: 987.11–987.36 m; **d** fine conglomerate and mudstone, scour surface, depth: 1019.2–1019.3 m; **e** grey argillaceous siltstone, shellfish fossils, well J2, depth: 991–991.1 m; **f** gray green mudstone, horizontal bedding, well A261, depth: 1017.36–1017.42 m

3.2.3.3 Underwater Distributary Interchannel Sand

Underwater distributary interchannel sand is sandwiched between underwater distributary channels and is locally connected with lake water. It consists mainly of grayish-green siltstone and argillaceous siltstone, with a small amount of fine sandstone. Sedimentary structures demonstrate parallel bedding, wavy bedding, and flaser bedding. The SP curve is flat and negative, and the RT curve is mostly low and flat and tooth-, finger-, or pinnacle-shaped occasionally, with RT > 3.5 W m

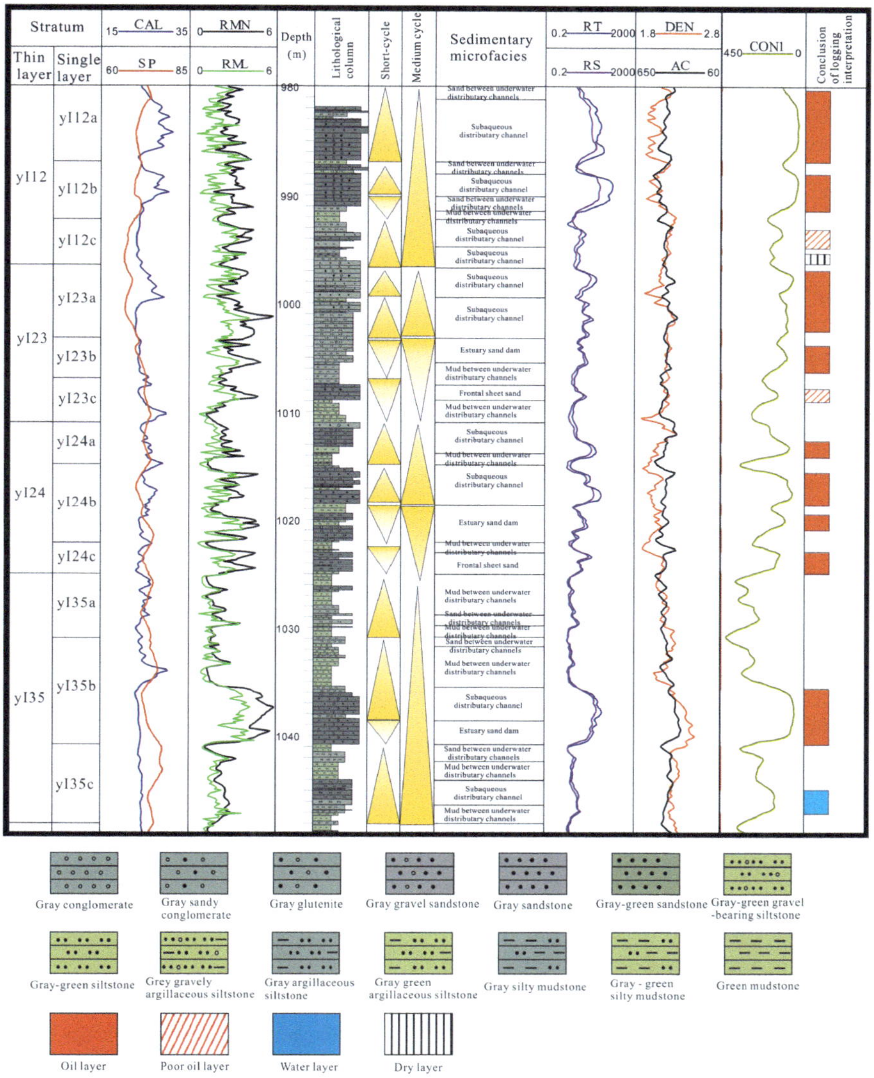

Fig. 3.14 The rock electrical characteristics of different sedimentary microfacies in Yulou oil layer of a certain area in the western depression of Liaohe Basin

(Fig. 3.14). This microfacies is mainly developed on the flanks of the underwater distributary channel and generally has small vertical thickness, small size, and poor reservoir properties. Nonetheless, it is also an important origin of the reservoir in the study area due to its high degree of development in the Yulou reservoir.

3.2.3.4 Underwater Distributary Interchannel Mud

Underwater distributary interchannel mud is located between underwater distributary channels and connected with lake water. It is mainly composed of grayish-green mudstone and silty mudstone, with a small amount of silty fine sandstone. Its sedimentary structures show horizontal bedding, wavy bedding, lenticular bedding, and convoluted bedding, suggesting a high strength of bioturbation. The SP curve shows a straight segment(s) parallel to the mudstone baseline locally, and the RT curve is mostly low and flat and tooth-, finger-, and pinnacle-shaped. This microfacies is developed locally in the Yulou reservoir and takes a majority part in nonreservoirs. It mainly appears in the form of a barrier/interlayer in the process of reservoir development (Fig. 3.14).

3.2.3.5 Delta-Front Sheet Sand

Delta-front sheet sand is fining upward, lithologically composed of sandstone and siltstone, with argillaceous rock, fine sandstone, and siltstone alternating with mudstone. It is well sorted and rounded, showing high maturity. Ripple cross bedding is observed in arenaceous rock, and ripple bedding, horizontal bedding, and bioturbation structure are developed in argillaceous rock. The SP and RT curves are finger- or pinnacle-shaped, with RT > 1.8 W·m. The ML (microlog) curve shows a low-amplitude funnel or jagged shape (Fig. 3.14). Delta-front sheet sand is generally developed at the head of the fan-delta front. Limited by the small area and other factors of the study area, front-sheet sand is rare in the Yulou reservoir and is observed locally in some sublayers near the S-E edge.

3.2.4 Characteristics of Sedimentary Microfacies

3.2.4.1 Sectional Characteristics

Multiple sections are established to clarify the distribution of sedimentary microfacies in the directions parallel and perpendicular to the origin (Fig. 3.15). Overall, the underwater distributary channel and underwater distributary interchannel sand are dominant, while the river-mouth bar and delta-front sheet sand are few. The small proportion of river-mouth bar mainly results from the frequent diversions of underwater distributary channels. The underwater distributary interchannel sand appears at channel flanks, and the interchannels are separated by mudstone, silty mudstone, and other fine-grained sediments. Multiple transgression-regression cycles are recorded from $yII_3{}^{6b}$ to $yI_1{}^{1a}$, and the Yulou reservoir as a whole shows a coarsening-upward trend. The underwater distributary channels are superposed and intersected laterally and distributed continuously locally. Sand bodies exhibit good connectivity in the direction parallel to the provenance. yI is roughly divided into the upper, middle, and lower parts. The sand bodies in the upper and lower parts are remarkably larger in size but less in number than those in the middle part. Sand bodies show obviously poor connectivity in the direction perpendicular to the provenance, which mainly results from the frequent diversions of underwater distributary channels. The river-mouth bar and delta-front sheet sand are more developed in yI than in yII, with more sand bodies in the former. The underwater distributary channels generally have a width range of 200–300 m and extend up to hundreds of meters, with numerous sand bodies of small size, revealing the typical sedimentary features of the fan-delta front. The underwater distributary interchannel sands occur as lamina between underwater distributary channel sand bodies. Laterally, underwater distributary channel sand bodies of different depositional periods or of the same depositional period are superposed and intersected, forming the framework sand bodies in the Yulou reservoir. The areas between channels or the superposed and intersected parts of channels tend to form seepage barriers because of varied reservoir properties, which may have great effects on steam flooding.

3.2.4.2 Plane Characteristics

Based on the well-based and sectional distributions of sedimentary microfacies and the sand thickness (isopach) maps, the plane distribution of sedimentary microfacies is mapped for 29 sublayers. For $yI_1{}^{1a}$ and $yI_1{}^{1b}$ (Fig. 3.16), the channels on the fan migrate constantly with the varied sediment supply in flood and drought periods. The channels at different parts of the fan form the framework sand bodies of the fan-delta front, which dominate the Yulou reservoir. These sand bodies are separated by fine-grained sediments such as underwater distributary interchannel mud. The underwater distributary channels appear in bands parallel to the origin, extending from the northwest to the southeast, and the underwater distributary interchannel sands are developed in narrow strips at channel flanks. The underwater distributary channels are superposed and intersected laterally and are continuous locally. The abovementioned distribution laws are consistent with the sectional characteristics of sedimentary microfacies, reflecting the constant diversions of the underwater distributary channel.

3.2.5 Influences of Sedimentary Features on Field Development

3.2.5.1 Sedimentary Texture

Sedimentary texture, which is the basis for the hydrocarbon occurrence in the reservoir, should be clarified first to understand how the sedimentary features of the reservoir affect oil and gas field development. The Yulou reservoir is composed of

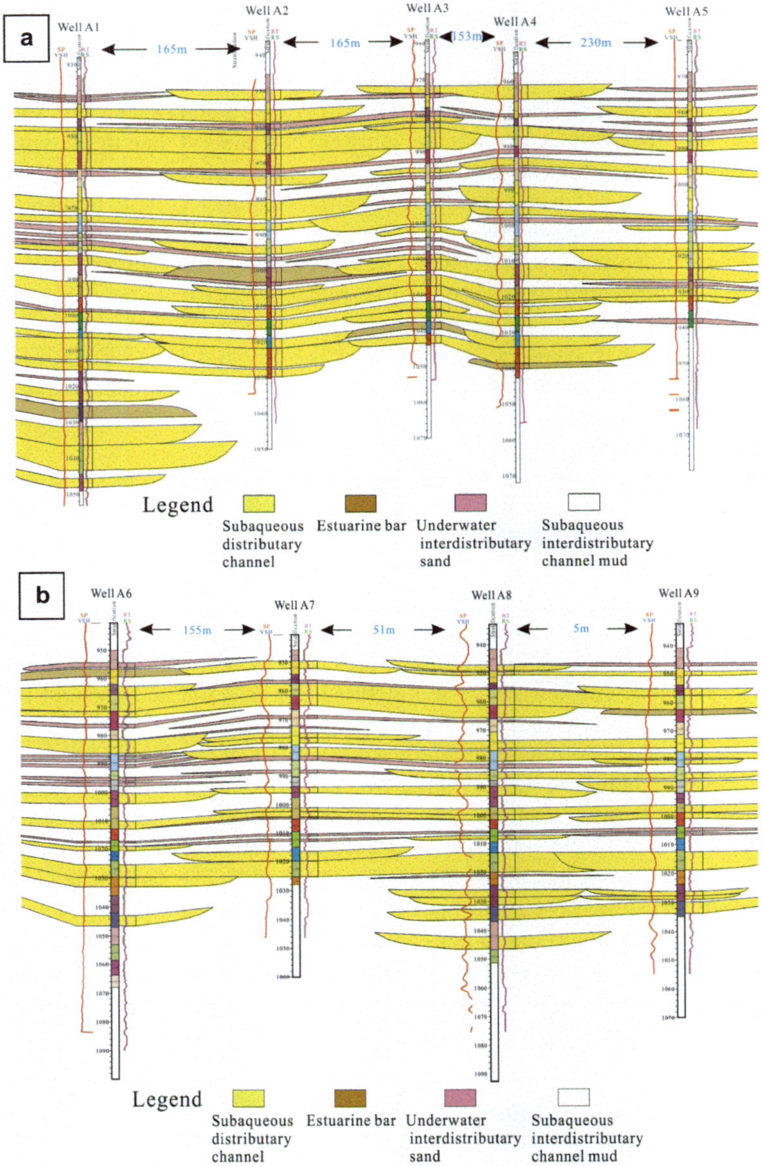

Fig. 3.15 Development characteristics of sedimentary microfacies profile of Yulou oil layer in a certain area of western sag of Liaohe Basin. **a** parallel provenance direction (NW–SE direction); **b** vertical provenance direction profile (NE–SW direction). *Note* Different colors in the layered column of single well column represent different single layer positions

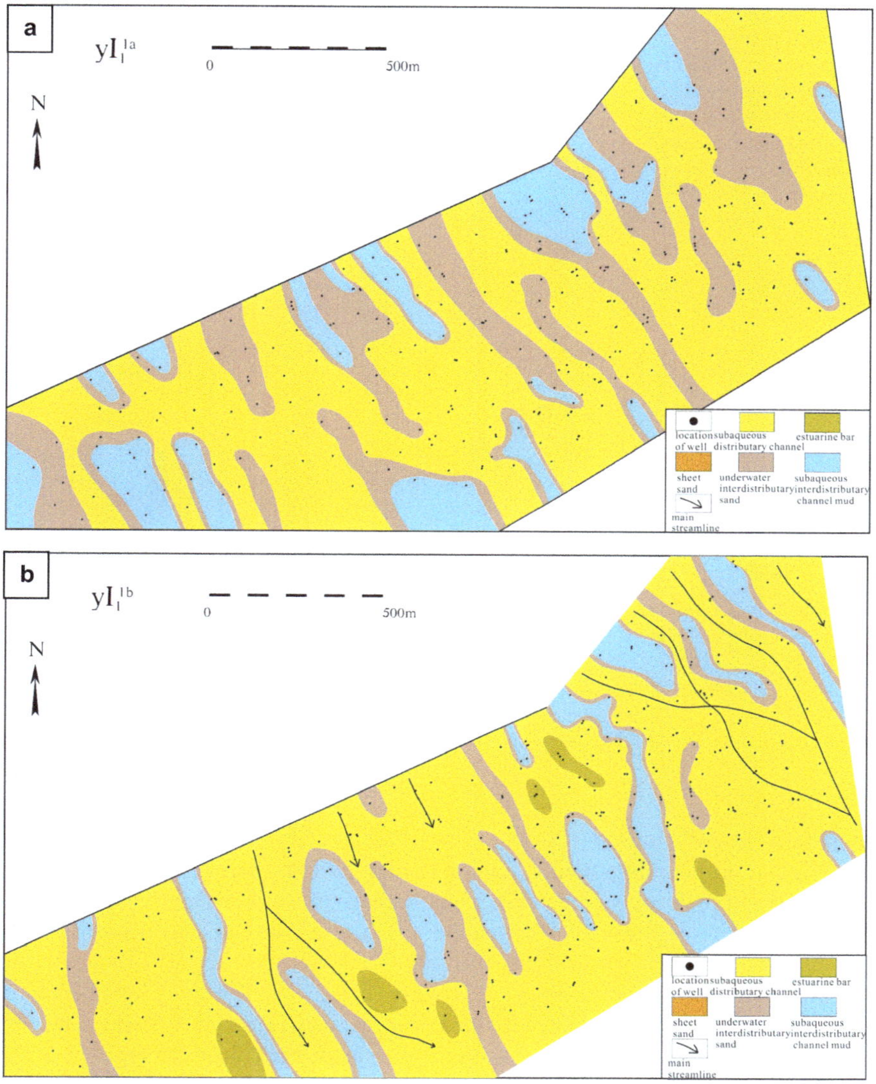

Fig. 3.16 Plane distribution map of sedimentary microfacies of Yulou oil layer in a certain area of western sag of Liaohe Basin. A-single layer yI1 1a sedimentary microfacies plane distribution map; b-single layer yI1 1b sedimentary microfacies plane distribution map

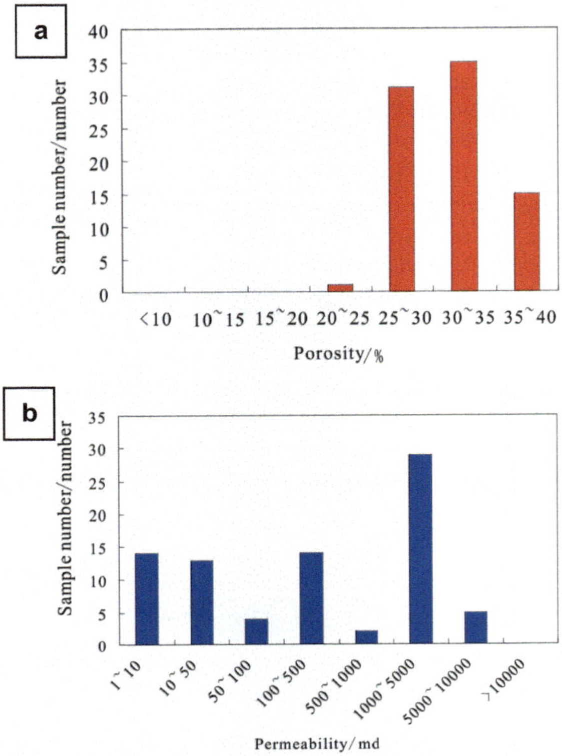

Fig. 3.17 Reservoir physical property characteristics of Yulou oil layer in a certain area of western sag in Liaohe basin. **a** Porosity characteristics; **b** permeability characteristics

fan-delta front sediments, which are coarse-grained and poorly sorted. However, it reflects good oil-bearing potential, owing to high porosity and permeability, which are contributed by loosely cemented grains because of weak diagenesis. According to the physical properties of the cores, the Yulou reservoir has a porosity of 25%–40% (avg. 31.25%) and permeability of 1–5000 mD (avg. 1829.3 mD), indicative of high-porosity and high-permeability reservoirs (Fig. 3.17). The study area develops positive, reverse, and composite rhythms, with positive rhythm in dominance. In the process of steam flooding, hot water in the steam front flows downward while moving horizontally due to the effect of gravity so that the heavy oil in the fine-grained sediments in the lower part of the reverse rhythm is recovered to enhance the total oil recovery. In general, rhythmic sediments have a very important influence on the performance of steam flooding.

3.2.5.2 Sedimentary *Microfacies or Sand Body Associations*

Sedimentary features influence and control the heavy oil thermal recovery. As shown in Fig. 3.18, the sedimentary microfacies of 29 sublayers in the Yulou reservoir are divided into lateral and vertical associations (Fig. 3.18). The lateral associations of sedimentary microfacies include lateral intersection/superposition of the underwater distributary channel and river-mouth bar, lateral intersection/superposition of the underwater distributary channels, and lateral intersection/superposition of the underwater distributary channel and underwater distributary interchannel. The vertical associations of sedimentary microfacies include vertical intersection/superposition of the underwater distributary channel and river-mouth bar, vertical intersection/superposition of underwater distributary channels, vertical superposition of the underwater distributary channel and underwater distributary interchannel, and vertical isolation of the underwater distributary channel. Among the lateral associations, the lateral intersection/superposition of underwater distributary channels and the lateral intersection/superposition of the underwater distributary channel and underwater distributary interchannel are most developed. Among the vertical associations, the vertical intersection/superposition of underwater distributary channels and the vertical superposition of the underwater distributary channel and underwater distributary interchannel are developed in 17 sublayers in yI (the upper Yulou reservoir), and the vertical isolation of the underwater distributary channel is developed in 12 sublayers in yII (the lower Yulou reservoir); the vertical intersection/superposition of the underwater distributary channel and river-mouth bar has not been observed in any sublayer.

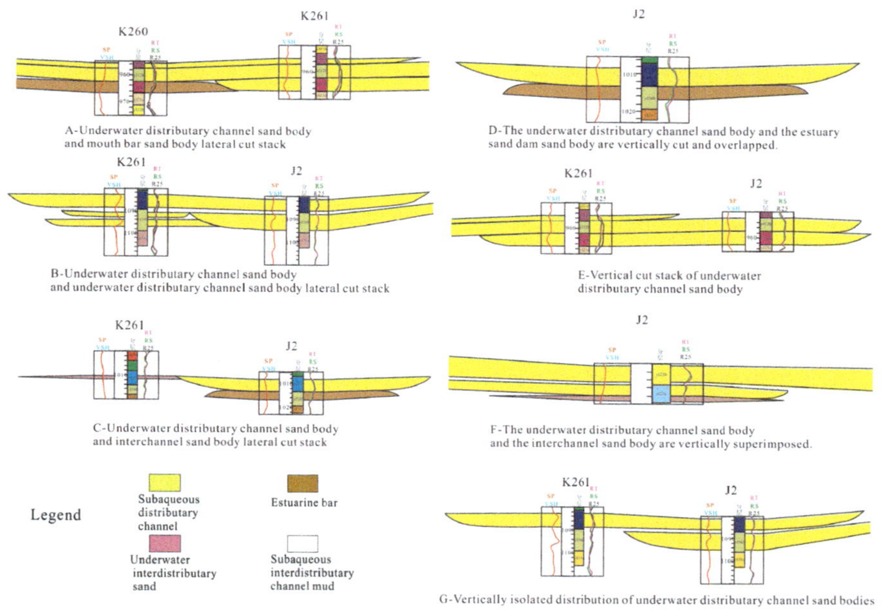

Fig. 3.18 Characteristics of single sand body combination mode of Yulou oil layer in a certain area of western depression of Liaohe Basin

Sedimentary microfacies associations play a very important role in controlling reservoir characteristics and oilfield development (Ma et al. 2008; Zhang et al. 2008a, b). Among lateral associations, for example, the lateral intersection/superposition of underwater distributary channels shows much better connectivity than the lateral intersection/superposition of the underwater distributary channel and underwater distributary interchannel. As the sand bodies are distributed in a large area horizontally and with a large thickness vertically, the steam injector and oil producers in a flooding unit are preferentially deployed in the reservoirs with the lateral intersection/superposition of underwater distributary channels to achieve the best injector/producer pairs for an optimal flooding effect. Among the vertical associations, the vertical superposition of underwater distributary channels demonstrates the best connectivity. Thus, the superposed sand bodies should be considered as a whole, and in this case, separate-layer steam injection is not recommended because steam channeling is likely to occur due to the absence of a barrier/interlayer. In the case of the vertical superposition of the underwater distributary channel and underwater distributary interchannel, separate-layer steam injection can be considered to facilitate the remaining oil recovery depending on the horizontal distribution and vertical thickness of interchannel sand bodies. Furthermore, the absorption of barriers/interlayers such as interchannel sand to steam heat energy should be considered to ensure the economic effectiveness of heavy oil recovery by steam flooding. In the case of the vertical isolation of the underwater distributary channel, the steam injector and oil producers must be deployed in the same sand body to achieve the best injector/producer pairs and production results. It illustrates the controls of microfacies associations on injector/producer connectivity according to the tracer tests (Fig. 3.19). In Fig. 3.19a, the sand bodies in yI_1^{1a}, yI_1^{1b}, and yI_1^{2a} between the steam injector K260 and the oil producer 27C2 present as the lateral superposition of the underwater distributary channels and vertical superposition of the underwater distributary channel and underwater distributary interchannel. The tracer was detected in 27C2 in the first week of the 3-month monitoring period after tracer injection in K260, suggesting good connectivity of the sand bodies. The sand bodies in yI_2^{3c}, yI_2^{4a}, and yI_2^{4b} between the steam injector K280 and the oil producer K29 present as isolation, lateral superposition of underwater distributary interchannel sand and underwater distributary channel, and lateral superposition of the river-mouth bar and underwater distributary channel. No tracer was detected in K29 in the 3-month monitoring period after tracer injection in K280, indicating poor connectivity of the sand bodies. These two examples further prove the control of microfacies associations on interwell connectivity.

3.2.5.3 Characteristics of Sedimentary Microfacies

Here, dynamic monitoring data of the steam front are used to analyze the influence of sedimentary microfacies on steam flooding. Two flooding units are selected as examples (Fig. 1.11), where Z1 and Z2 are steam injectors, and hollow and solid circles represent producers without and with the steam front monitored, respectively.

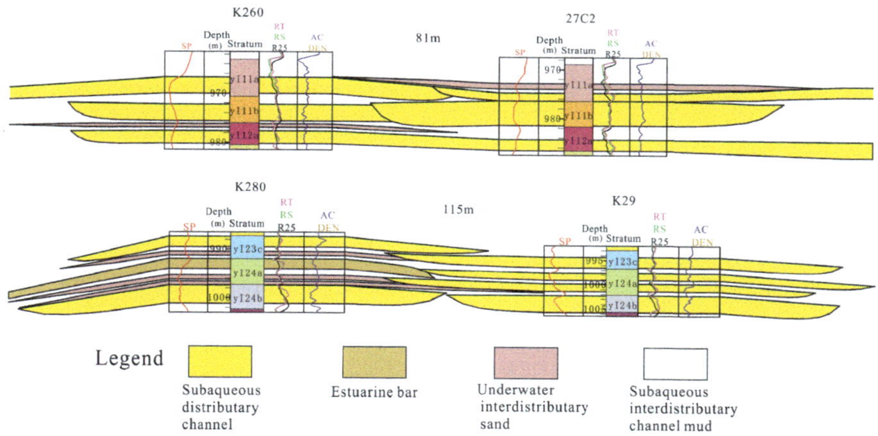

Fig. 3.19 The influence of single sand body combination mode on interwell connectivity in Yulou oil layer in a certain area of western sag of Liaohe Basin (Based on tracer data)

Fronts of steam injected into Z1 and Z2 are monitored to diagnose the connectivity of reservoirs between steam injectors and oil producers. Note that the reservoirs in the study area are loose and weakly reformed by diagenesis, so the reservoir properties are mainly controlled by sedimentary microfacies. In the flooding unit with the steam injector Z1, B1–B4 and B8 are preferential response wells, B5–B7 are subpreferential response wells, and B9–B11 are nonresponse wells. The producers with obvious responses are mainly arranged in the direction parallel to the provenance, indicating that the steam front is apparently controlled by sedimentary microfacies and mainly advances along the main stream line. B9–B11 are not affected by steam because they are far away from Z1. Changes in main channels owing to the frequent diversions of underwater distributary channels deteriorate the reservoir connectivity. Similar situations are seen in the flooding unit with the steam injector Z2.

These findings show that the injector and producer should be ideally drilled in the same sedimentary facies belt without sedimentary facies variation between the two wells to ensure the best injector/producer configuration. Moreover, the best flooding results can be achieved when the injector and producer are located along the main stream line of the same underwater distributary channel. In the case of the underwater distributary channel at the fan-delta front, the steam front preferentially advances along the main stream line of each distributary channel. Several distributary channels may connect laterally to form a large-scale oil-bearing sand body, but the resulting lateral connectivity is worse than the connectivity of the same channel. Therefore, it is recommended to deploy the injector and producers in one flooding unit in one sand body of the distributary channel to achieve the flooding performance.

3.3 Geological Genetic Analysis of Barriers/Interlayers in Fan-Delta Front Reservoirs

"Barrier" or "interlayer" in reservoir research refers to the lowly permeable or impermeable rock zone relative to the oil zone. A barrier is defined as an impermeable layer in the reservoir that can prevent or control fluid motion. It has greatly varied thickness, up to tens of meters, and a long stretching distance. An interlayer refers to a relatively impermeable layer distributed in sandstone. It has unstable distribution, small thickness, generally ranging from only a few centimeters to tens of centimeters, and short stretching distance (Li 2007; Wu et al. 2010; Zhang et al. 2011). The core of the geological study of reservoirs is heterogeneity (Ye et al. 2004; Qin 2014). The barrier/interlayer, as an essential part in heterogeneity research, has a critical impact on the effective reservoir development and remaining oil recovery and has always been one of the hot issues in field development (Lin et al. 1997; Yan et al. 2008; Wang et al. 2009; Yong et al. 2010; Hu et al. 2010; Zhang et al. 2013; Gao et al. 2013a, b). Previous studies dealt with barriers/interlayers mainly using sedimentology, diagenesis, geostatistics, well logging geology, geological modeling, and other methods and covered origin analysis and classification, identification based on petroelectric characteristics, and impact on field development. They targeted clastic reservoirs of the fluvial, alluvial fan, delta, and other facies, as well as carbonate reservoirs. Generally, they focused on conventional water flooding reservoirs. However, research on thermally recovered heavy oil reservoirs, especially the influence of the barrier/interlayer on steam huff-and-puff and steam flooding, has been rarely reported. Fan-delta front reservoirs are characterized by short sediment transport distance owing to the proximity to the origin, a larger number of small-sized sand bodies, frequent diversions of underwater distributary channels (Galloway et al. 1983; Mei 1994; Jiang et al. 2003; Zhu 2008a, b), and the presence of barriers/interlayers. In this section, the barriers/interlayers in thermally recovered heavy oil reservoirs are discussed based on geological, logging, core, seismic, and test data to provide a geological basis for the conversion to steam flooding and enhancement of oil recovery. The information of the study area and the target horizon are detailed in Sect. 3.2.

3.3.1 Genetic Classification and Identification of Barriers/Interlayers

Based on geological, logging, core, and test data, through geological setting investigation, comparative analysis of logging curves, detailed core description, and careful observation and statistical analysis of test data, the barriers/interlayers in the Yulou reservoir are classified into sedimentary, diagenetic, and composite origins (Table 3.3, Fig. 3.20). The barriers/interlayers of sedimentary origin are mainly controlled by sedimentation, mostly of lake-flooding mudstone and underwater distributary interchannel mud. They are lithologically composed of siltstone, argillaceous siltstone, silty mudstone, mudstone, and other fine-grained sediments. They are usually found in two forms. One is the argillaceous layer formed in the drought period during lake flooding. It is generally thick and often appears as the barrier (Chen et al. 2015a, b). It shows continuous strong reflectance on the seismic profile, corresponding to the top and bottom boundaries of the Yulou reservoir and the boundary between yI and yII. The other is the argillaceous layer between sandy laminae formed by the changes in hydrodynamic conditions resulting from diversions of underwater distributary channels. It is generally thin and dominantly occurs in the form of an interlayer (Chen et al. 2015a, b). The barriers/interlayers of sedimentary origin cannot be identified by seismic data but only by well data. They generally reflect low and flat SP curves, hole enlargement on CAL curves, and low RT values (Fig. 3.21). They are developed widely in both yI and yII, with large vertical thickness. The barriers/interlayers of diagenetic origin also exist in two forms. One is the calcareous interlayer formed by atmospheric acid water. Its formation and distribution are related to faults. Atmospheric water is likely to flow along faults, and the calcareous interlayer can be observed near faults in the study area. The other is the calcareous interlayer formed by the acid fluid generated by property variation of $NaHCO_3$ type formation water, water–rock interactions, and fault movement in the study area. It shows distinct high values on the logging curves (Fig. 3.22; Chen et al. 2015a, b). It is generally approximately 1-m thick, and mainly distributed in yII, but rarely in yI (Chen et al. 2015a, b). The barriers/interlayers of the composite origin are formed by the joint action of sedimentation and diagenesis and exist in two forms. One is the fine-grained materials or even coal streaks originating from the carbonization of plant debris in clastic rocks, which exist in the bedding interface of sandstone or siltstone or in mudstone. It has generally small thickness in the range of millimeters to centimeters. The other is the cemented calcareous sediments where the calcium is derived from fossil shells by replacement in the diagenetic process of snail-rich formations. It has generally large thickness in the range of centimeters to decimeters. The barriers/interlayers of composite origin can be identified from the observation of core samples collected from coring wells but are difficult to recognize on logging curves due to their small thickness (commonly < 1 m) (Fig. 3.23). They often occur as thin interlayers in formations. In practice, large-scale barriers/interlayers are mainly delineated by tracking seismic events and identifying large sets of mudstones on logging curves, and smaller barriers/interlayers are primarily identified from the results of fine-logging interpretations.

3.3.2 Spatial Distribution of Barriers/Interlayers

3.3.2.1 Sectional Distribution of Barriers/Interlayers

The Yulou reservoir is dominated by barriers/interlayers of sedimentary origin. So, the identification of the barrier/interlayer, in a sense, signifies the delineation of sand bodies because the barrier/interlayer is exactly developed in nonsand horizons.

Table 3.3 Classification of barriers/interlayers in the Yulou reservoir in an area of the Liaohe Basin's West Sag

Classification by origin		Identification data	Vertical thickness	Degree of development
Sedimentation	Diversions of underwater distributary channels, change in hydrodynamic conditions	Core observation and logging data	Centimeters to meters	Developed
	Lake-flooding mudstone	Core observation, logging data, and seismic data	Centimeters to meters	Developed
Diagenesis	Atmospheric acidic water	Core observation and logging data	Centimeters to meters	Undeveloped
	Formation water	Core observation and logging data	Centimeters to meters	Undeveloped
Composite	Carbonation of plant debris in clastic rocks	Core observation data	Millimeters to centimeters	Developed occasionally
	Calcification of clastic rocks containing abundant snail fossils during diagenesis	Core observation and logging data	Centimeters to decimeters	Developed occasionally

The scale and distribution of interwell barriers/interlayers are predicted based on statistical analysis and interwell parameter prediction of the sand body architectures in the reservoir. On the section parallel to the source direction, barriers/interlayers are generally stable, relatively continuous, and thick. They are composed mainly of lake-flooding mudstone and silty and argillaceous sediments between underwater distributary channels of different sedimentary periods and the same sedimentary period. The existence of underwater distributary interchannel sand also complicates the spatial distribution of barriers/interlayers. The frequent diversions of channels lead to the spatially superposed sand bodies. The newer channels transform the older sand bodies or fine-grained sediments, resulting in the thickness variation of barriers/interlayers. On the section perpendicular to the source direction, barriers/interlayers exhibit significantly greater thickness variation than that on the section parallel to the source direction because of the poorer connectivity of sand bodies, as well as larger thickness and better continuity. The barriers/interlayers have significantly decreased thickness in the locally superposed sand bodies of the underwater distributary channel and show large thickness and good continuity in the sand bodies of smaller channel.

3.3.2.2 Plane Distribution of Barriers

According to the analysis of barriers/interlayers in different wells and the sectional distribution of barriers/interlayers, the plane distribution of barriers in 29 sublayers is mapped. It illustrates the plane distribution of barriers from yI_3^{5c} to yI_3^{6a} (Fig. 3.24).

Fig. 3.20 Core characteristics of interlayer development in Yulou oil layer in a certain area of western sag of Liaohe Basin. **a** W1 well, 999.32–999.36 m, gray-black mudstone; **b** W2 well, 1040.8–1041 m, gray-green mudstone; **c** W3 well, carbonate mineral, 965.72 m, sandstone, 6000× ; **d** W4 well, 1033.84–1033.94 m, medium sandstone, siliceous cementation; **e** W5 well, 1041.12–1041.16, gray mudstone, carbonized plant debris; **f** W4 well, 1003–1003.1 m, silty mudstone, cephalopod fossils (snails)

It is seen that, evidentially controlled by sedimentary facies, the barriers distribute roughly in the NW–SE direction. They are continuous locally with large thickness but are absent in some areas where sand bodies of different periods are superposed due to diversions of distributary channels and scouring of older underwater distributary channels by younger ones. Barriers in the rest of the sublayers show similar distribution characteristics but slight differences in the plane distribution maps. According to the statistics, barriers are relatively developed in the Yulou reservoir, with a maximum thickness up to 18.66 m (Table 3.4). However, they show remarkable differences. Direct superimposition of sand bodies in the absence of the barrier can be observed at each sublayer interface. The barriers are mostly thicker than 2.5 m. The physical simulations, numerical simulations, and production practice in the Liaohe Oilfield demonstrate that a barrier with a thickness of > 1 m can play a favorable role in sealing during steam flooding (Qiu et al. 1996; Liu 1997, 1998a, b). It is found that barriers/interlayers act as effective seals in most parts of the study area. As mainly formed in the lake-flooding period and locally controlled by ancient landform, the barriers are characterized by generally stable and locally varied thickness.

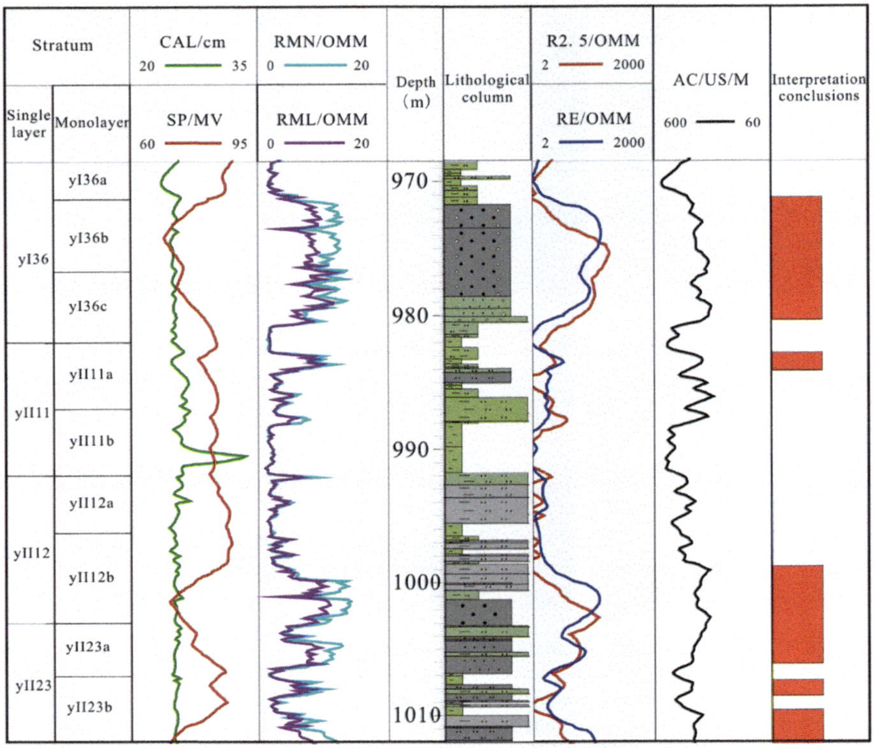

Fig. 3.21 Lithology-electricity characteristic of sedimentogenesis interlayer of some well in aimed stratum in study area (Depth: 988–991.5m)

3.3.2.3 Plane Distribution of Interlayers

The interlayer has a significantly smaller spatial extension and vertical thickness than the barrier. In this book, the plane distribution of interlayers is described according to the interlayer density maps. Interlayer density refers to the ratio of the total thickness of interlayers in sand bodies to the total thickness of sand bodies (including interlayers) (Wu et al. 1998), which is expressed as a percentile. Interlayer density maps are drawn for 29 sublayers. In yI_3^{6a}, for example, interlayers are more apparently controlled by diversions of underwater distributary channels than barriers (Fig. 3.25). Interlayers are distributed intermittently in most parts of the study area and continuously only in a small area locally, which impacts fluid motion in the reservoir. Similar to barriers, interlayers also distribute roughly in the NW–SE direction under the control of the source direction. At the sublayer level, it is found that over 50% of the sublayers contain only one sand body; hence, the interlayer is considered not developed in general. At the layer level, the interlayer is mostly identified as a barrier.

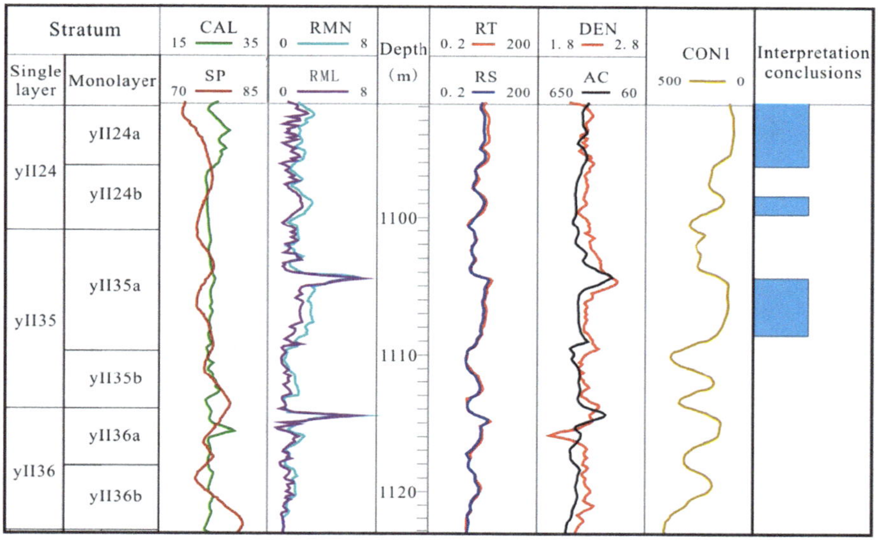

Fig. 3.22 Lithology-electricity characteristic of diagenesis causes interlayer of some well in aimed stratum in study area (Depth: 988–991.5m) (Chen et al. 2015b)

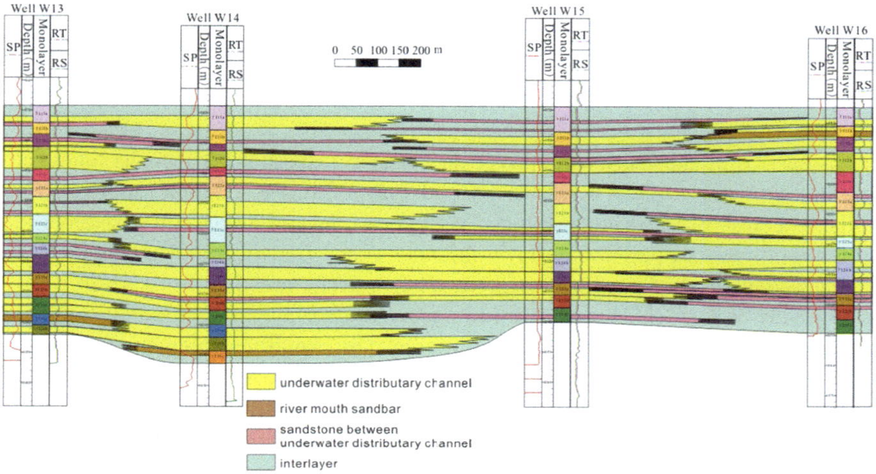

Fig. 3.23 section map of interlayer development characteristics in aim layer in study area. The mudstone interlayer is widely distributed in space and plays an important role in controlling the connectivity of different types of sedimentary sand bodies. Mudstone interlayer is controlled by sedimentation and its thickness varies greatly in space

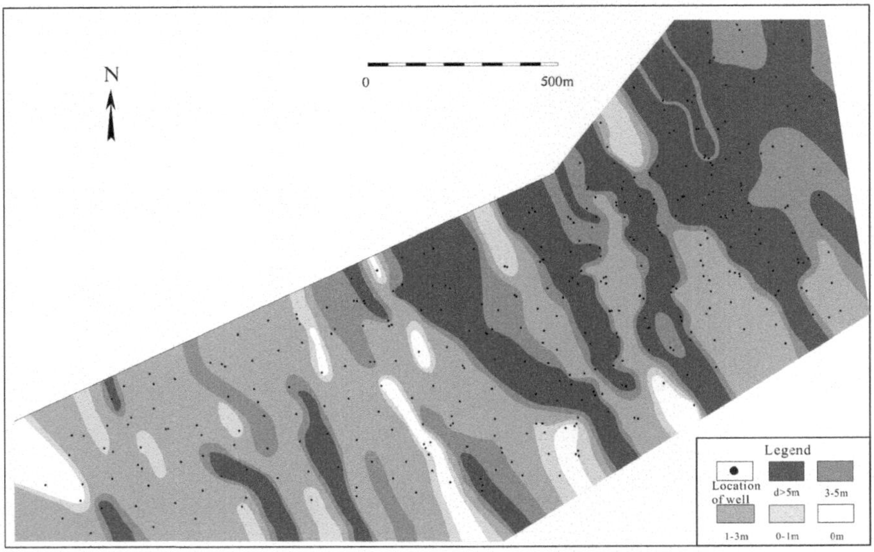

Fig. 3.24 The development characteristics of interlayer between single layer yI35c–yI36a of Yulou oil layer in a certain area of western sag of Liaohe Basin

3.3.3 Influence of the Barrier/Interlayer on Heavy Oil Thermal Recovery

The influence of the barrier/interlayer on heavy oil thermal recovery is analyzed. The flooding unit with W6 (Table 3.5 and Fig. 3.26), for example, includes five wells, where yI oil zones are developed with good connectivity. A comparison of the formation thickness, barrier/interlayer thickness, oil production, and steam injection indicates that the barriers/interlayers in all five wells are mainly of sedimentary origin. Wells W7 and W8 have the thinnest barriers/interlayers and the smallest steam injection but achieve the highest oil production. The situation is opposite for the rest of the three wells. This is largely due to the gradually decreasing efficiency of steam flooding and the worsening injector-producer relationship with the increase in barrier/interlayer thickness. The production practice of this flooding unit has confirmed the importance of barriers/interlayers in controlling and influencing the production of thermally recovered heavy oil reservoirs. Similar findings are observed in the other eight flooding units in the study area. It is concluded that barriers/interlayers have positive and negative impacts on heavy oil thermal recovery by steam flooding.

(1) Negative impacts

(a) The irregularly distributed mudstone zigzags the channels for fluid flow in the reservoir, complicating the injector-producer relationship. As shown in Fig. 3.11, barriers/interlayers separate the sand bodies deposited in different

Table 3.4 Statistical characterization of interlayer thickness between different single layers in aimed stratum in study area

Numerical value	Interlayer name						
	yI11a–yI11b	yI11b–yI12a	yI12a–yI12b	yI12b–yI12c	yI12c–yI23a	yI123a–yI23b	yI23b–yI23c
Maximum	13	13.78	9.81	12.16	13.34	10.02	9.4
Minimum	0	0	0	0	0	0	0
Average	3.04	3.06	1.6	2.18	3.84	3.61	1.52
Numerical value	yI23c–yI24a	yI24a–yI24b	yI24b–yI24c	yI24c–yI35a	yI35a–yI35b	yI35b–yI35c	yI35c–yI36a
Maximum	12.5	9.99	9.15	9.74	11.71	10.38	15.63
Minimum	0	0	0	0	0	0	0
Average	2.68	2.22	2.31	3.31	2.07	2.29	3.69
Numerical value	yI36a–yI36b	yI36b–yI36c	yI36c–yII11a	yII11a–yII11b	yII11b–yII12a	yII12a–yII12b	yII12b–yII23a
Maximum	12.2	11.47	11.5	15	13.76	17	15.14
Minimum	0	0	0	0	0	0	0
Average	2.32	2.11	5.2	3.99	4.93	5.26	6.31
Numerical value	yII23a–yII23b	yII23b–yII24a	yII24a–yII24b	yII24b–yII35a	yII35a–yII35b	yII35b–yII36a	yII36a–yII36b
Maximum	14.97	14.35	13.54	12.71	15	14.8	18.66
Minimum	0	0	0	0	0	0	0
Average	5.83	5.4	4.35	4.96	3.51	3.48	6.24

The average thickness of interlayer is more than 2.5 m for most single layers. The average and maximum thickness of the interlayer is 18.66 m. The stable distribution of separation interlayer plays an important role in the thermal recovery and development of heavy oil

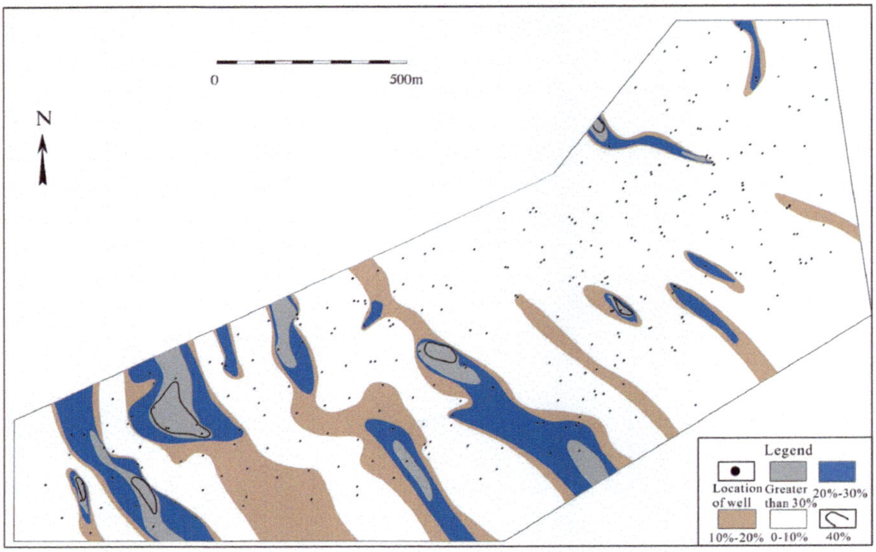

Fig. 3.25 Distribution characteristics of interlayer density in single layer yI36a of Yulou oil layer in a certain area of western sag of Liaohe Basin

Table 3.5 Correlation table of stratum thickness and interlayer thickness and production data in aimed stratum (yI11a–yII12a) in study area

Items	Well name				
	Well7	Well6	Well9	Well8	Well10
Strata thickness	101.091	106.986	97.793	104.478	106.37
Interlayer thickness	43.76	47.78	61.43	33.51	57.49
Oil production	4420.6	339.2	2033.9	4077.3	2454.8
Steam injection volume	4100	29,647	5427	2385	5095

periods on the section. As a result, the pathway for lateral migration of injected steam varies with barriers/interlayers, making the steam motion flexural complicate the injector-producer relationship. In Figs. 3.12 and 3.13, barriers/interlayers l deteriorate the connectivity of reservoirs on the plane and separate them into bands. The distribution of barriers/interlayers is irregular under the control of underwater distributary channels; hence, the plane motion of the injected steam is complicated, and the injector-producer relationship is finally complicated.

(b) As a large amount of heat is absorbed by barriers/interlayers in steam flooding, the effective heat transfer is difficult to realize in reservoirs for oil displacement, resulting in a worse economic effect of steam flooding. In terms of the steam flooding mechanism, steam injection is primarily

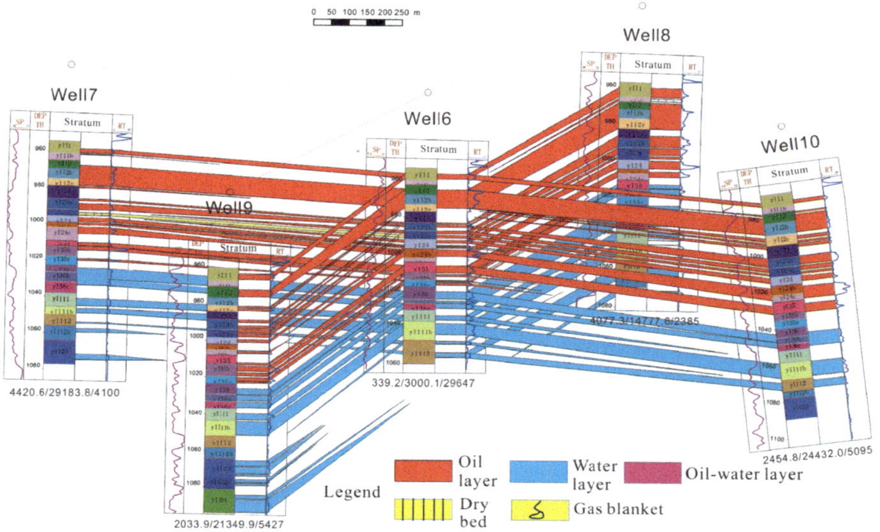

Fig. 3.26 Reservoir development characteristic pattern of Well6 well group in aim layer in study area. *Note* the digital representation of the underground department: oil production/water yield / steam injection. The well group production data in the figure fully shows that the separation and interlayer has a very important influence on the thermal recovery and development of heavy oil. We should pay more attention to the research of interlayer in the process of oilfield development

for changing the reservoir properties and enhancing the fluidity of crude oil by heating the reservoirs. However, if the oil-free barriers/interlayers that are close to the oil zones are relatively developed, they will absorb a lot of heat, resulting in serious heat loss to induce low usage of steam. This is also a unique phenomenon in heavy oil thermal recovery by steam flooding, which is different from conventional water flooding, so it should be considered carefully.

(c) When reducing well spacing and adjusting the injector-producer relationship, the impact of barriers/interlayers on production should be minimized to ensure favorable injector-producer pairs. The presence of barriers/interlayers between the injector and producers will make the steam motion preclude, thereby damaging the injector-producer relationship.

(2) Positive impacts

Vertically, at the zones where barriers are relatively developed, separate-layer steam injection can be adopted to minimize heat loss during steam flooding and thus improve development efficiency (Chen et al. 2015a, b). At the zones where barriers are not developed between sublayers, only commingled steam injection is recommended. The existence of barriers/interlayers impacts the motion path of injected steam in formations and reduces the influence of gravity on heavy oil thermal recovery; consequently, the injected steam can move farther laterally along the formation, improving the efficiency of heavy oil thermal recovery.

Overall, the existence of barriers/interlayers is more negative than positive for steam flooding. Therefore, the influence of barriers/interlayers should be fully considered when the development mode is converted to steam flooding after steam huff-and-puff to improve the thermal recovery performance of heavy oil reservoirs of the fan-delta front.

3.4 Geological Genesis Analysis and Classified Evaluation of Fan-Delta Front Reservoirs

After a detailed analysis of the factors affecting the origin of fan-delta front reservoirs, the reservoirs in the Yulou reservoir are qualitatively evaluated by SPSS cluster analysis with selected characteristic parameters and the fine logging interpretation results of 400 wells. Close integration of qualitative genesis analysis and quantitative classified evaluation fully guarantees the rationality of selected parameters and the reliability of evaluation results to provide a geological basis for the conversion of development mode for fan-delta front reservoirs. The information on the study area and the target horizon are detailed in Sect. 3.2.

3.4.1 Characteristics of Reservoirs

3.4.1.1 Lithology

Based on the observation and description of cores with a total length of 668 m collected from six coring wells (Fig. 3.27), combined with the grain-size analysis of 233 samples, it is indicated that the Yulou reservoir has various types of rocks, including fine conglomerate, glutenite, coarse sandstone, medium sandstone, fine sandstone, siltstone, argillaceous siltstone, silty mudstone, and mudstone, which are dominantly gray, grayish black, and grayish green, mostly containing sands and gravels, with high-shale content. Medium–fine sandstones are predominant.

3.4.1.2 Sedimentary Microfacies

Using the geological, logging, and core data, the Yulou reservoir is determined as fan-delta front sediments by analyzing the single-well, sectional, and plane distributions of sedimentary facies, which is subdivided into underwater distributary channel, river-mouth bar, underwater distributary interchannel sand, underwater distributary

Fig. 3.27 Lithological core photo characteristics of Yulou oil layer in a certain area of western sag of Liaohe Basin. **a** A10,944.73–944.78 m, gray-brown fine conglomerate; **b** A2,945.39–945.53 m, grayish-brown fine sandstone; **c** A2,948–948.1 m, siltstone; **d** A22,1040.8-1041 m, mudstone

interchannel mud, and delta-front sheet sand microfacies (Fig. 3.28). The underwater distributary channel and river-mouth bar are predominant.

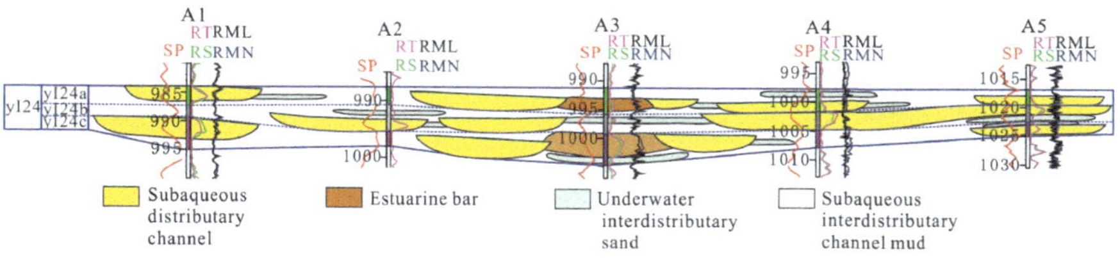

Fig. 3.28 Development characteristics of sedimentary microfacies profile of Yulou oil layer in a certain area of western sag of Liaohe Basin

3.4.1.3 Pore Structure

The pore structure of a reservoir refers to the geometry, size, distribution of pores and throats in rocks, and the connection between them. In-depth analysis of the pore structure for revealing the reservoir architecture is meaningful for oil and gas exploration and development (Chen et al. 2013a, b, c). The intrusive mercury curve of the Yulou reservoir appears to be inclined to the bottom left (Fig. 3.29), indicating that the pores and throats are well sorted, coarse skewed, and well developed. Through core analysis and thin-section observation, the pore structure of the Yulou reservoir is believed to include primary pores and secondary pores, which can be subcategorized into intergranular pore, intragranular pore, matrix micropore, cleavage fracture, intergranular dissolution pore, intragranular solution pore, moldic pore, extra-large intergranular dissolution pore, structural fracture, and dissolution fracture. Each pore type is reflected specifically on the thin section (Fig. 3.30). Generally, intergranular pores and intergranular dissolution pores are predominant.

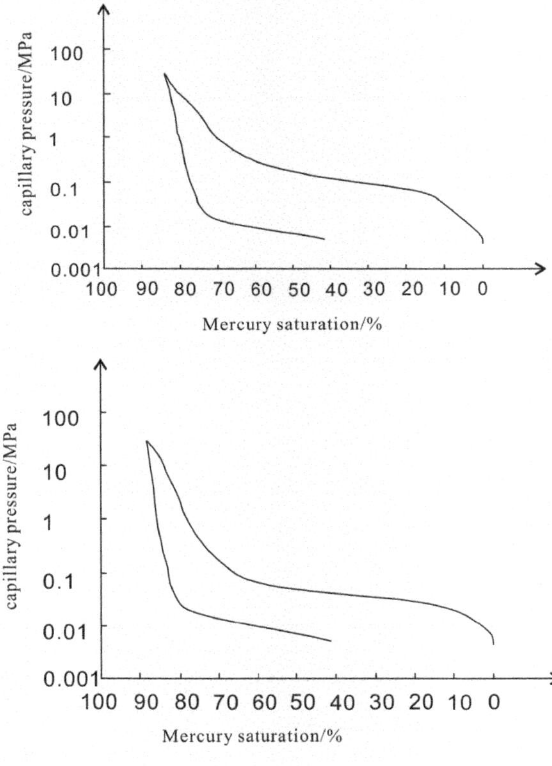

Fig. 3.29 Characteristics of mercury injection curve of well J1 in Yulou oil layer in a certain area of western sag of Liaohe Basin. **a** yI, depth: 923.18 m; **b** yII, depth: 1001.99 m

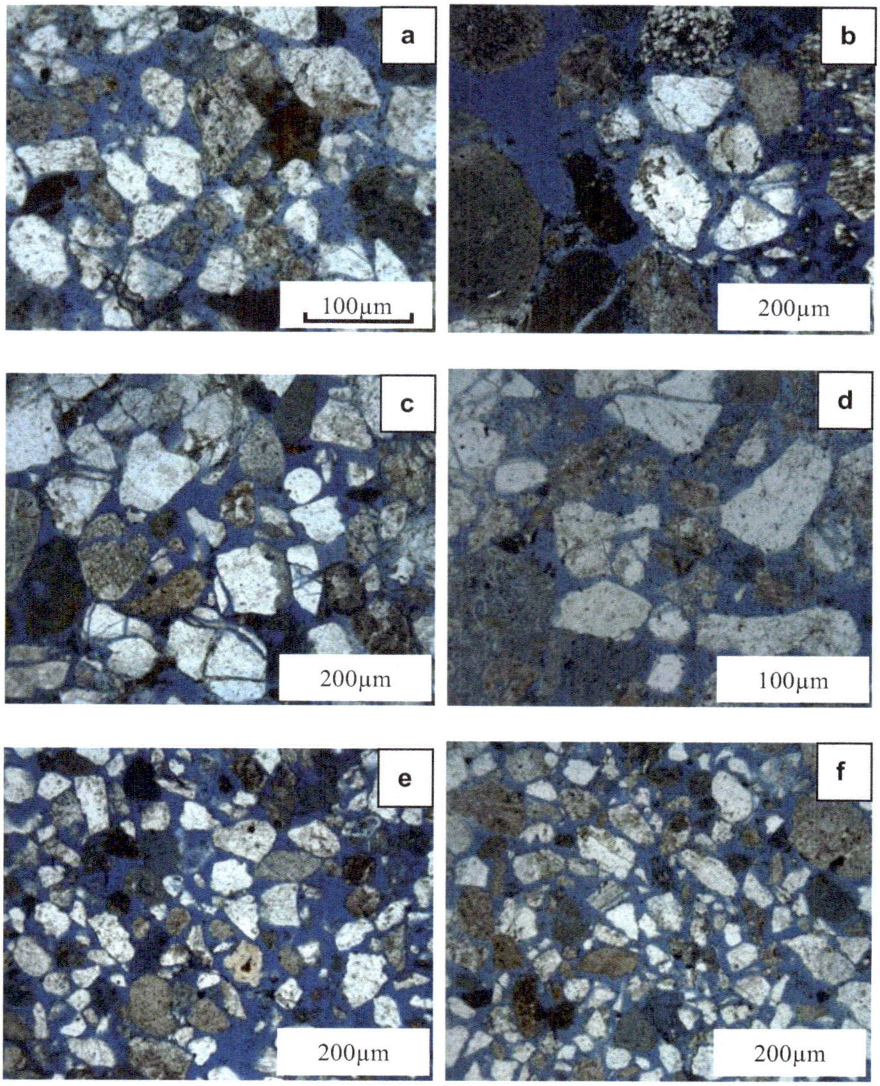

Fig. 3.30 Pore structure characteristics of Yulou reservoir in a certain area of western sag of Liaohe Basin. **a** A2 well, intergranular pore, depth: 945.3 m; **b** A2 well, extra-large dissolution pore, depth: 965.72 m; **c** A2 well, structural fracture, depth: 974.56 m; **d** A2 well, matrix micropore, depth: 960.32 m, 50×; **e** A2 well, mold hole, depth: 1022.72 m; **f** A2 well, cleavage fracture, depth: 1005.72 m

3.4.1.4 Physical Properties

According to the analysis and test results of 150 cores, the Yulou reservoir exhibits a porosity in the range of 25–40%, with an average of 31.25%, and permeability of 1–5000 mD, with an average of 1829.3 mD, indicative of high-porosity and high-permeability reservoirs.

3.4.1.5 Diagenesis

Diagenesis is closely related to reservoir properties (Chen et al. 2013a, b, c). The pore structure of the Yulou reservoir is also affected by diagenesis, mainly including cementation and dissolution (Fig. 3.31). Cementation includes carbonate cementation and clay mineral cementation, in which clay mineral cementation is more common.

3.4.1.6 Reservoir Heterogeneity

Reservoir heterogeneity is one of the major factors controlling ground oil, gas, and water migration and hydrocarbon recovery. The study of reservoir heterogeneity is of great significance to the effective development of oil and gas reservoirs (Chen et al. 2012). By analyzing the single-well sedimentary rhythms, it is found that the Yulou reservoir is controlled by sedimentation and has strong heterogeneity vertically. Three sedimentary rhythms have been identified in the Yulou reservoir, including positive rhythm, reverse rhythm, and composite rhythm, with the positive rhythm and composite rhythm in dominance. The vertical variation in sedimentary rhythms results in the strong reservoir heterogeneity. By analyzing the plane distribution of the permeability variation coefficient of yI sublayers, it is demonstrated that the reservoir heterogeneity is apparently controlled by sedimentary microfacies. $yI_1{}^{2a}$, $yI_1{}^{2b}$, and $yI_1{}^{2c}$ show the weakest reservoir heterogeneity, where the area with the permeability variation coefficient < 0.5 accounts for approximately half of the total study area. $yI_2{}^{3a}$, $yI_2{}^{4a}$, $yI_2{}^{4b}$, and $yI_2{}^{4c}$ have the second weakest reservoir heterogeneity, where the area with the permeability variation coefficient < 0.5 accounts for approximately one-third of the total study area. The reservoir heterogeneity of the rest of sublayers is generally strong, where the area with the permeability variation coefficient < 0.5 is distributed sporadically. This is consistent with the distributions of porosity and permeability.

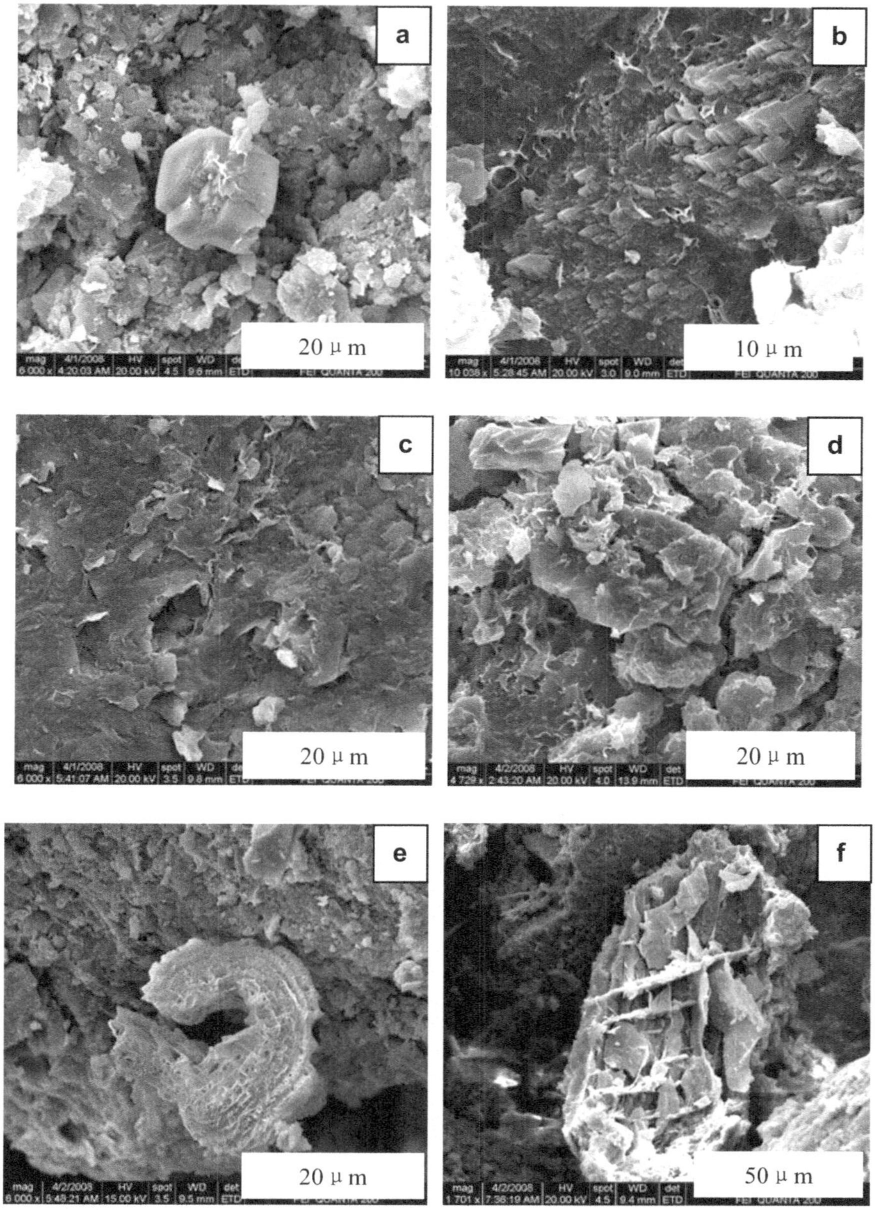

Fig. 3.31 Diagenesis characteristics of Yulou oil layer in a certain area of western sag of Liaohe Basin. **a** A2 well, carbonate minerals, 965.72 m, sandstone; **b** A2 well, feldspar secondary enlargement up to grade I, 979.32 m, sandstone; **c** A2 well, grain surface flake illite clay, 984.35 m, sandstone; **d** A2 well, montmorillonite clay, 989.18 m, sandstone; **e** A2 well, kaolinite clay, vermicular, 1005.72 m, sandstone; **f** A2 well, particle dissolution phenomenon, 1022.72 m, sandstone

3.4.2 Quantitative Evaluation of Reservoirs

3.4.2.1 Parameter Selection

Parameter selection is crucial to reservoir evaluation. Numerous factors impact reservoir properties; thus, researchers may select different evaluation parameters. More parameters are not always better. Too many parameters may make some evaluation factors be considered repeatedly and the major controlling factors of reservoir quality difficult to recognize. Based on the genetic analysis of reservoirs, the parameters of reservoir evaluation are selected. Tectonism primarily affects the reservoir properties through its generated fractures, which can be reflected by permeability. Sedimentation plays a crucial role in controlling the reservoir properties. The frequent diversions of underwater distributary channels further aggravate the reservoir heterogeneity in space. This can be reflected by the permeability variation coefficient. Shale content is also a very important indicator because it can indicate whether the Yulou reservoir is of underwater distributary channel or interchannel sediments. Areas with high porosity mainly correspond to underwater distributary channels or river-mouth bars, while those with low porosity mostly correspond to underwater distributary interchannel mud. Thus, porosity is also an important indicator. As a comprehensive embodiment of the reservoir properties, the reservoir thickness reflects lithology, lithofacies, diagenesis, and other factors to a certain extent. After comprehensive consideration, five parameters, that is, shale content, porosity, permeability, reservoir thickness, and permeability variation coefficient, are selected for the quantitative evaluation of the Yulou reservoir.

3.4.2.2 Evaluation Method

Q-type clustering, one major species of the hierarchical clustering in SPSS, is used in the procedure of clustering. The so-called Q-type clustering is performed to cluster according to the characteristics of variables. All cases with similar characteristics are classified into one category (Cai et al. 2006). Clustering analysis for reservoir classification by shale content, porosity, permeability, thickness, and permeability variation coefficient belongs to Q-type clustering. In clustering, the first step is to normalize the five variables, allowing the value comparison and mathematical operation to be possible for the variables. The clustering analysis methods in SPSS software, including the Pearson correlation, Chebychev, Minkowski, Block, and Ward's method, are compared according to the principle that effective classification can be realized and the classification results using the five parameters have the same change trend and conform to the current development status. Finally, Ward's method is selected. An effective classification indicates that none of the classified reservoir categories has a significantly smaller number of results than other categories (difference in order of magnitude) to reflect the rationality of the selected classification parameters, calculation methods, and classification results. The consistent change trend of the classification results using the five parameters refers to the classification results that ensure that the samples with large porosity, high permeability, and small shale content are classified into the same class, favorable class I or class II reservoir, and the samples with small porosity, permeability, and shale content are classified into the same class, poor class III or class IV reservoir. In this way, the influence of the large-permeability and small-porosity points and other singularities resulting from microfractures can be avoided in the classified evaluation results of reservoirs. The conformation to the current development status refers to that the favorable class I and class II reservoirs identified by the classified evaluation have good development performance at present and the poor class III and class IV reservoirs correspond to the parts with poor production results currently. The application of Ward's method is determined by repeated tests, and researchers' experience in clustering analysis is also necessary. The selected parameters and calculation methods may be different depending on areas and data. Finally, the classification results are obtained by partition and merging according to distance and relationship (Table 3.6, Fig. 3.32).

3.4.2.3 Evaluation Results

The Yulou reservoirs are divided into class I, class II, class III, and class IV. With favorable physical properties, class I and class II reservoirs are the main development targets. Class III and class IV reservoirs have poor physical properties and are difficult to develop economically by existing techniques. These reservoirs are apparently controlled by sedimentary facies. Class I and class II reservoirs are mostly located in the underwater distributary channel and river-mouth bar, while class III and class IV reservoirs are in the underwater distributary interchannel sand or delta-front sheet sand.

3.4.2.4 Reservoir Characteristics

The plane distribution of different classes of reservoirs in each sublayer is analyzed. The sedimentary factors are determined as a leading role in the process of reservoir forming. The four classes of reservoirs generally have a banding distribution along the NW–SE direction, which is basically consistent with the distribution direction of sedimentary microfacies. Vertically, the distribution horizons of reservoir properties vary with migration and diversion of the main underwater distributary channels of different sedimentary periods at the fan-delta front. In yI, class I and class II reservoirs are mainly developed in yI_1^{2a}, yI_1^{2b}, yI_1^{2c}, yI_2^{3a}, yI_2^{3b}, yI_2^{3c}, yI_2^{4a}, yI_2^{4b}, and yI_2^{4c}, of which yI_1^{2a}, yI_1^{2b}, and yI_1^{2c} have the best reservoir properties.

Table 3.6 Statistical characteristics of reservoir classification and evaluation parameters of Yulou oil layer in a certain area of western sag of Liaohe Basin

Reservoir classification	Parameter														
	Porosity/%			Permeability/μm^2			Argillaceous content/%			Effective thickness/m			Coefficient of variation		
	Maximum	Minimum	Average	Maximum	Minimum	Average	Maximum	Minimum	Average	Maximum	Minimum	Average	Maximum	Minimum	Average
Type I reservoir	39.000	35.170	38.590	6.057	4.618	5.520	39.980	0.660	17.650	7.260	0.000	2.870	0.522	0.000	0.194
Type I I reservoir	38.240	23.730	36.580	4.615	2.709	3.594	44.990	1.240	21.520	8.450	0.000	2.920	1.030	0.147	0.635
Type III reservoir	37.110	23.140	34.350	2.748	1.001	1.751	49.970	1.160	24.710	10.980	0.000	2.560	1.780	0.000	0.917
Type IV reservoir	35.030	6.180	30.050	1.346	0.000	0.355	54.840	0.310	27.300	8.660	0.000	1.450	6.012	0.000	0.921

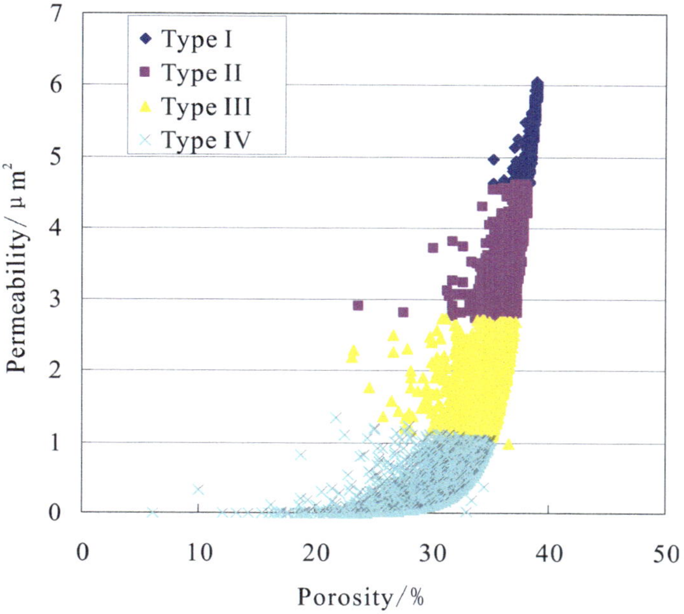

Fig. 3.32 Porosity–permeability relationship characteristics of single-layer reservoir evaluation and classification results of Yulou oil layer in a certain area of western sag of Liaohe Basin

3.4.2.5 Rationality and Verification of Evaluation Results

The reservoir evaluation by clustering analysis is a systematic process. Its rationality is ensured in the following ways. First, appropriate parameters that fully reflect the factors influencing the reservoir properties are selected. Second, optimal calculation methods and data normalization algorithms are chosen to ensure that the correct-judging ratio of the classified evaluation results that come with the software is more than 85%. Third, a certain number proportion is guaranteed for every class to eliminate the influence of a few singular values. Fourth, the relationships of parameters are compared for classification results to ensure that all parameters play a role in the clustering procedure to achieve a comprehensive and rational clustering analysis. Fifth, good correlation and consistency of reservoir evaluation results and sedimentary microfacies are shown by comparing their plane distributions (Fig. 3.33). This also indicates the rationality of the evaluation results and fully reflects the control of sedimentary factors on reservoir properties.

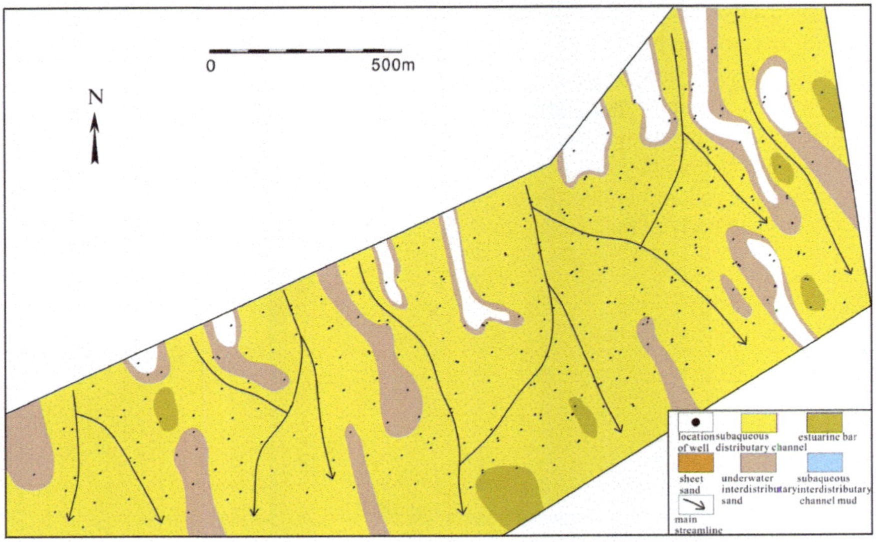

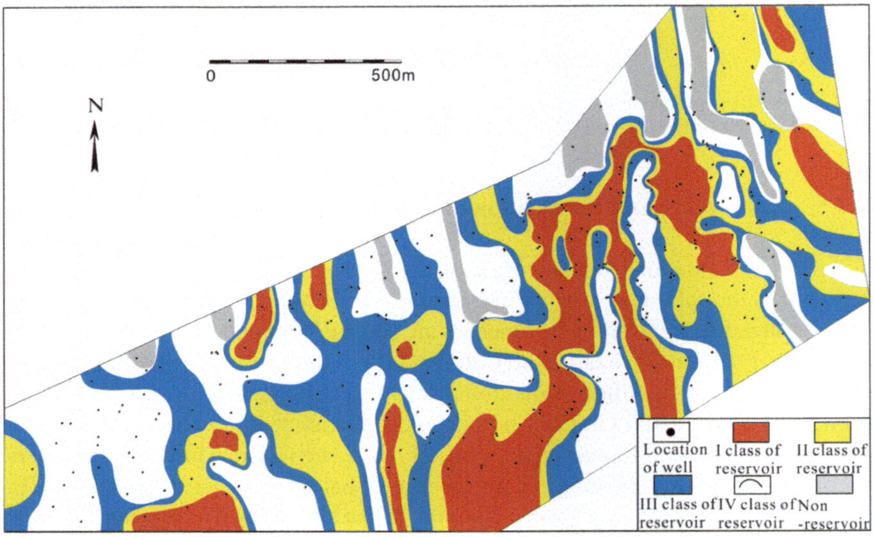

Fig. 3.33 Comparison of plane distribution characteristics of sedimentary microfacies and reservoir classification evaluation results of Yulou oil layer in a certain area of western sag of Liaohe Basin. **a** Single-layer yI12b sedimentary microfacies plane distribution map; **b** Plane distribution map of single layer yI12b reservoir evaluation results

3.4.3 *Influence of Classified Evaluation Results on the Conversion of Thermal Recovery Mode*

The results of the reservoir evaluation can provide a solid geological basis for the conversion to steam flooding after steam huff-and-puff for heavy oil reservoirs of fan-delta front. As the reservoir is the material foundation of steam flooding, its properties directly determine the effect of steam flooding.

The following should be considered during steam flooding.

(1) Depending on the results of reservoir evaluation, the steam injector and oil producers should be deployed in zones with consistent or similar physical properties to achieve a good injector-producer relationship. In the study area, steam injectors and oil producers should be drilled in class I and class II reservoirs, and the reservoirs between these wells should be the same class, distributed continuously and stably without any change.
(2) The injector-producer spacing should be reasonably designed according to the reservoir properties to prevent steam channeling. Well pattern and well spacing should be customized for different classes of reservoirs when the development mode is converted.
(3) For yI and yII, the reservoir evaluation results of sublayers should be fully considered to allow steam flooding to work effectively in as many sublayers as possible.

Under the current development situations, most sublayers are commingled for production considering the economic effectiveness of thermal recovery by steam flooding, while different sublayers have varied reservoir properties. Therefore, it is necessary to ensure that an injection-production system covers the sublayers in class I and class II reservoirs to achieve the optimal development performance.

3.5 Summary

The Lower Karamay Formation in the northwestern margin of the Junggar Basin is mainly composed of medium sandstone, sandy conglomerate, and fine conglomerate. It contains grains generally larger than 2 mm and poorly rounded, predominantly appearing subrounded to subangular. The log-probability grain-size distribution curves show that many rock types are dominated by coarse-grained rolling and saltation populations but rarely fine-grained suspension populations. Influenced by the current scour, the saltation population of the samples collected from the braided flow belt is divided into two subpopulations, which show as two intersecting line segments and slight difference in median value and degree of sorting. The C-M and scatter diagrams of structural parameters indicate that the Lower Karamay Formation is dominated by traction-current deposits and controlled by retrograding sequences, where the depositional characteristics of the traction current are increasingly noticeable from bottom to top.

The grain-size analysis shows that the Lower Karamay Formation in the northwestern margin of the Junggar Basin is of typical alluvial-fan conglomerate reservoirs characterized by coarse grain size and poor sorting. Grain-size analysis is widely used in investigating sedimentary environments. As one of the powerful tools for sedimentary facies analysis, grain-size analysis is very useful. However, it has limitations and multiplicity of solution. Therefore, it should be closely integrated with and cross-verified by other relevant data and methods to draw objective and true conclusions.

The Yulou reservoir in the Western Sag, the Liaohe Basin can be subdivided into 29 sublayers, corresponding to 29 short-term base-level cycles. It is mainly composed of medium–fine sandstone and develops trough cross bedding, tabular cross bedding, wavy bedding, flaser bedding, parallel bedding, and horizontal bedding. The color of the mudstone suggests a reducing environment, and abundant cephalopod fossils (snails) are observed in the mudstone and siltstone. The log-probability grain-size distribution curves show two-segment and three-segment patterns, indicating sediments transported by traction, saltation, and suspension before deposition. Through a comprehensive analysis of lithology, bedding, mudstone color, fossils, and grain size, the Yulou reservoir is determined as the sediments of the underwater distributary channel of the fan-delta front.

Five microfacies have been identified in the Yulou reservoir, namely, underwater distributary channel, river-mouth bar, delta-front sheet sand, underwater distributary interchannel sand, and underwater distributary interchannel mud. Underwater distributary channel microfacies are predominant. The underwater distributary interchannel mud is developed between underwater distributary channels and is the main component of nonreservoirs, generally appearing in the form of a barrier/interlayer.

The underwater distributary channel sand bodies of the Yulou reservoir appear as bands in the NW–SE direction parallel to the provenance. The underwater distributary interchannel sand is developed in channel flanks, and the channels are separated by fine-grained sediments such as silty mud and mudstone. Sublayers $yII_3{}^{6b}$–$yI_1{}^{1a}$ present an upward-coarsening trend as a whole. Generally, the channels range in width from 200 to 300 m and extend up to hundreds of meters, with a large number and small size of sand bodies. With the change in sediment supply in flood and drought periods, channels in the fan migrate constantly.

The sedimentary features of the Yulou reservoir influence and control the heavy oil thermal recovery. The sedimentary microfacies associations can be divided into lateral and vertical associations, which are subdivided into seven types. It is indicated that the injector and producers should be deployed in the same sedimentary facies belt without sedimentary facies change between wells to ensure the best injector-producer relationship. Moreover, the flooding effect will be the best when the injector and producers are located on the mainstream line of the same underwater distributary channel. In the case of the underwater distributary channel at the fan-delta front, the steam front preferentially advances along the main stream line of each distributary channel.

The frequent diversions of channels lead to the spatially superposed sand bodies in the Yulou reservoir, resulting in the thickness variation of barriers/interlayers. Apparently controlled by sedimentary facies, barriers are distributed in the near NW–SE direction. Barriers are relatively developed in the Yulou reservoir, with a maximum thickness of up to 18.66 m. Overall, the barriers are thicker than 2.5 m and can act as effective seals. As mainly formed in the lake-flooding period and controlled by ancient landform, the barriers/interlayers are characterized by generally stable and locally varied thickness. Interlayers have significantly smaller spatial extension and vertical thickness than barriers but similar distribution laws. The barriers/interlayers formed by the underwater distributary interchannel mud are in dominance, and those formed by the calcification of underwater distributary interchannel sand are rare and only observed in yII.

The production practice of the heavy oil thermal recovery of the Yulou reservoir shows the importance of barriers/interlayers in controlling and influencing the thermal recovery performance of a heavy oil reservoir. The negative impacts are as follows: (a) The irregularly distributed mudstone zigzags the channels for fluid flow in the reservoir, complicating the injector-producer relationship. (b) As a large amount of heat is absorbed by barriers/interlayers in steam flooding, effective heat transfer is difficult to realize in reservoirs for oil displacement, resulting in a worse economic effect of steam flooding. (c) When reducing well spacing and adjusting the injector-producer relationship, the impact of barriers/interlayers on production should be minimized to ensure favorable injector-producer pairs. The positive impacts are as follows: (a) At the zones where barriers are relatively developed, separate-layer steam injection can be adopted to minimize heat loss during steam flooding and thus improve the development efficiency. (b) At the zones where barriers are not developed between sublayers, only combined steam injection is recommended. Overall, the existence of barriers/interlayers is more negative than positive for steam flooding. As the steam huff-and-puff and steam flooding have especially significant differences from conventional water flooding, research on barriers/interlayers is necessary.

The Yulou reservoir is mainly composed of medium–fine sandstones and is dominated by the underwater distributary channel and river-mouth bar. The pores and throats are well developed, coarse skewed, and well sorted. Generally, they are intergranular pores and intergranular dissolution pores. The reservoirs exhibit an average porosity of 31.25% and average permeability of 1829.3 mD, indicative of high-porosity and high-permeability reservoirs. The main diagenetic processes are cementation and dissolution. Three sedimentary rhythms have been identified, namely, positive rhythm, reverse rhythm, and composite rhythm, of which positive rhythm and composite rhythm are in dominance. The reservoir heterogeneity is strong as a whole.

Five parameters, that is, shale content, porosity, permeability, reservoir thickness, and permeability variation, which can fully reflect the influencing factors of the reservoir, are selected for the quantitative evaluation of the reservoirs in the Yulou reservoir. According to the evaluation results, the reservoirs are divided into four classes, of which class I and class II have good physical properties and act as the main development targets. The development of reservoirs of different classes is mainly controlled by sedimentary facies. Class I and class II reservoirs are mostly located in the underwater distributary channel and river-mouth bar, while class III and class IV reservoirs are in the underwater distributary interchannel sand or delta-front sheet sand.

The assurance and verification of the correctness and rationality of reservoir evaluation results are essential content in the reservoir evaluation. First, appropriate parameters that fully reflect the factors influencing the reservoir properties are selected. Second, optimal calculation methods and data normalization algorithms are chosen to ensure that the correct-judging ratio of the classified evaluation results that come with the software is more than 85%. The rationality of the reservoir evaluation results can be verified by comparing the plane distributions of reservoir evaluation results and sedimentary microfacies.

References

Bao Z, Lixin Z, Yong W, et al. The main control factors of sandbody reservoir development in rift-subsidence lake basin: an example from the Paleogene West Sag of Liaohe Basin. Geoscience. 2009;23(4):676–82.

Cai J, Yu H, Zhu Z, et al. SPSS statistical analysis examples selected. Beijing: Tsinghua University Press; 2006.

Chen H, Yushuang Z, Qingyin Li, et al. Sedimentary microfacies in Chang 6 oil-bearing formation of Yanchang formation of Xinghe District in Ansai Oilfield. J Northwest Univ (Nat Sci Edn). 2006;36(2):295–300.

Chen F, Ping L, Xingyang Z, et al. Characteristics of grain size of delta sandbody framework in Yanchang Formation Upper Triassic, North Shaanxi. Acta Sediementologica Sinca. 2010a;28(1):58–67.

Chen H, Xiaomin Z, Gongcheng Z, et al. Characteristics of migration system of Paleogene Lingshui Formation in the deep water area in Qiongdongnan Basin. Acta Geol Sin. 2010b;84(1):138–48.

Chen H, Xiaomin Z, Gongcheng Z, et al. Seismic facies in a deepwater area of a marine faulted basin: deepwater area of the Paleogene Lingshui formation in the Qingdongnan Basin. Acta Geologica Sinca. 2012;86(2):473–83.

Chen H, Chen C, Shuxian L, et al. Research advances on reservoir pores. Nat Gas Geosci. 2013a;24(2):227–37.

Chen H, Ping J, Danfeng Z, et al. Pore structure of volcanic reservoir classified and distribution characteristics evaluated: taking member 1 of Yingcheng formation in Xudong Area in Songliao Basin as an example. J Central South Univ (Sci Technol). 2013b;44(4):1453–63.

Chen H, Chunyan L, Jing Z, et al. Research advances of reservoir diageneses. Petrol Geol Oilfield Dev Daqing. 2013c;32(2):1–9.

Chen H, Zhao Y, Gao X, et al. High-resolution sequence stratigraphy and its application in the fine-scale stratigraphic correlation of the Yulou Reservoir in the West Depression of the Songliao Basin. J Stratigr. 2014;38(3):317–323.

Chen H, Shuxian L, Peng J, et al. Flow barrier characteristics of Thermal recovery of heavy oil reservoir: taking Yulou oil-bearing sets in some experimental area in the south end of westward slope of West depression in Liaohe Basin as an example. Acta Sedimentol Sin. 2015a;33(3):616–24.

Chen H, Lei X, Chang L, et al. Classified evaluation of the geological bodies in Yulou Oil Reservoirs of Liaohe Basin. Petrol Geol Oilfield Dev Daqing. 2015b;34(5):45–51.

Chen H, Hu Y, Shi C. Single factor characterization and quantitative evaluation of volcanic rock reservoirs in songliao basin. Beijing: Petroleum Industry Press; 2016a.
Chen H, Chengfang S, Chen C. Problems with the fine description of reservoirs. Pet Geol Exp. 2016b;38(5):569–76.
Fan S, Yang S, Su N, et al. Carved up and discerned of reservoir structure of interface of fan delta deposits- Taking Dujiatai oil bearing sets in Jin99 area in Huanxiling Oilfield in Liaohe Basin as an example. J Xi'an Shiyou Univ (Nat Sci Edn). 2015;30(1):7–13.
Galloway WE, Hobday DK. Terrigenous clastic depositional systems—applications to petroleum, coal, and uranium exploration. New York: Springer; 1983.
Gao L, Jun M, Tao Y, et al. Application of seismic attribute analysis technique to the identification of mudstone interbeds. Lithol Reserv. 2013a;25(4):101–5.
Gao Y, Wei S, Xin Z. Microscopic Geologic origin for Ultra-low permeability reservoir of Yanchang Groups, Ordos Basin. J Jilin Univ (Earth Sci Edn). 2013b;43(1):13–9.
Ge D, Liu Y, Liu X, et al. Application of grain size analysis in tight sandstone reservoir and sedimentary environment evaluation. Spec Oil Gas Reserv 2018;25(1):41–45, 72.
Hu W, Chao C, Wei W, et al. The study of the m acroscopic heterogeneity of reservoir: taking the Heidim iao reservoir of Da 208 block in Jilin Oilfield as an example. Chin J Geol. 2010;45(2):466–75.
Jiang Z. Sedimentology. Beijing: Petroleum Industry Press; 2003, p. 270–282.
Li Y, Jianmin L. Reservoir development geology. Beijing: Petroleum Industry Press; 2007. p. 34–5.
Li M, Jiafu Qi, Hengmao T, et al. Cenozoic fault structure and hydrocarbon accumulation in Western Sag, Liaohe Depression. Pet Explor Dev. 2010b;37(3):281–8.
Li G, Xu H, Lu Y, et al. High-resolution sequence stratigraphy for alluvial fan on lower Karamay formation in 8th zone of northwestern Junggar Basin. J Central South Univ (Sci Technol), 2010a;41(3):1124–1131.
Lin C, Lianhua H, Chunmei D, et al. Geostatistics applied to the identification of interbeds, with Sha-3 Member of Western Liaohe Depression Taken as an example. Exp Petrol Geol. 1997;19(3):245–51.
Lin Yu, Shenghe Wu, Dali Y, et al. Fine anatomizing reservoir architecture of fan-delta front: a case study on Dujiatai Reservoir in Shu2-6-6 Block, Liaohe Oilfield. Nat Gas Geosci. 2013;24(2):335–44.
Liu W. Thermal recovery engineering of heavy oil reservoir with steam flooding. Beijing: Petroleum Industry Press; 1997.
Liu H, Wu S, Yan P. Fracture research the latest progress and research advances in the fracture of the South China Sea. Nanhai Yanjiu Yu Kaifa; 1998a, p. 35–42.
Liu W. The development models of heavy oil reservoirs by thermal recovery. Beijing: Petroleum Industry Press; 1998b.
Ma S, Lv G, Yan B, et al. Research on three-dimensional heterogeneous model of channel sandbody controlled by architecture. Earth Sci Front. 2008;15(1):057–064.
Mei Z. Sedimentary facies and paleogeography Rebuild. Xi'an: Northwest University Press; 1994.
Pan F, Chunming L, Yanli Li, et al. Sediments grain-size characteristics and environmental evolution of core SE2 in southern bank of Qiangtang River since the Late Quaternary. J Palaeogeogr. 2011;13(2):236–44.
Qin R, Cao G, Wu Y. Review of the study of groundwater flow and solute transport in heterogeneous aquifer. Adv Earth Sci. 2014;29(1):30–41.
Qiu Y, Chen Z. Reservoir description. Beijing: Petroleum Industry Press; 1996.
Song Z, Jinlin Y, Lingli P, et al. Researches about favorable sedimentary belt of conglomerate reservoir by using grain-size analysis. Petrol Geol Recov Eff. 2005;12(6):16–8.
Sun S. Sedimentary characteristics and controlling factors of Xibaqian fan of Shahejie Formation, West Depression, Liaohe Basin. J Palaeogeogr. 2001;3(2):92–98.
Tang Y, Xu J, Qu J, et al. Fan: delta group charaeteristics and its distribution of the Triassic Baikouquan reservoirs in Mahu Sag of Junggar Basin. Xinjiang Petrol Geol. 2014;35(6):628–635.

Wang G, Shaochun Y, Feiyan L, et al. Hierarchical structure of barrier beds and interbeds in Braided River reservoirs. Nat Gas Geosci. 2009;20(3):278–383.
Wu S. Reservoir characterization & modeling. Beijing: Petroleum Industry Press; 2010, p. 157–161.
Wu S, Qihua X. Petroleum reservoir geology. Beijing: Petroleum Industry Press; 1998, p. 55–104.
Xinghe Yu, Daojian Z, Jianjun G, et al. Depositional facies and models of deep burial strata of the Shahejie formation in the Eastern and Western Depressions of Liaohe Oil Field. J Palaeogeogr. 1999;1(3):40–9.
Yan Y, Tianxiang D, et al. Identification and inter-well prediction of interbeds in thick oil layer. Lithol Reserv. 2008;20(2):127–31.
Yang T, Yingcheng C, Yanzhong W, et al. Genesis of high-quality reservoirs of fan delta front in lower part of the fourth member of Shahej ie Formation in Bonan Subsag. Earth Sci J China Univ Geosci. 2015;40(12):2067–80.
Ye S, Wu J, Xue Y. Application of multiscale finite element method to three dimensional groundwater flow problems in heterogeneous porous media. Adv Earth Sci. 2004;19(3):437–442.
Yong W, Jianhua Z. Outcrop characteristics of fan dalta of lake basin and its relationship with oil gas. Petrol Geol Recov Eff. 2010;17(3):6–11.
Yong Z, Yangsuo ZT, et al. Features and distribution of insulating layers of Bashijiqike Formation in Dalaoba area, Tarim Basin, China. J Chengdu Univ Technol (Sc Technol Edn). 2010;37(1):50–54.
Yu X, Qu J, Tan C, et al. Conglomerate lithofaeies and origin models of fan deltas of Baikouquan Formation in Mahu Sag, Junggar Basin. Xinjiang Petrol Geol. 2014;35(6):619–627.
Yuan H, Wang L, Yu Y, et al. Review of sedimentary grain size analysis methods. J Jilin Univ (Earth Sci Edn), 2019;49(2):380–393.
Zhang J, Jun X. Development geology. Beijing: Petroleum Industry Press; 2011. p. 60–5.
Zhang S-M, Tianyou P, Yulin Z. Application of quantitative analysis to Taiyuan Formation's and Shanxi Formation's Sedimentation Conditions in Boshan Area. J Shandong Univ Sci Technol (Nat Sci). 2003;22(4):45–8.
Zhang Pu, Jianqiang C, Mingzhong T, et al. Application 0fsediments grain-size analysis in Quaternary environmental research and its stratum compartmentalization and contrast function in Xiamen island, Fujian province. J Earth Sci Environ. 2005;27(1):88–94.
Zhang P, Chunhui S, Yongbiao Y, et al. The significance and establishment of discriminant function with grain size of stable lacustrine sediment and eolian loess. Acta Sedimentol Sin. 2008a;26(3):501–7.
Zhang Q, Bao Z, Song X, et al. Hierarchical division and origin of single sand bodies in Fuyu oil layer. Fuyu oilfield. Petrol Explor Dev 2008b;35(2):157–163.
Zhang C, Rui Z, Taiju Y, et al. Advances in Fan Deltaic sedimentology. Xinjiang Petrol Geol. 2015a;36(3):362–8.
Zhang W, Tao W, Chang He, et al. Structural characteristics and genetic analysis of Tongcheng fault in Jinhu sag. J China Univ Petrol. 2015b;39(5):18–26.
Zhang J, Yu X, Zhang T, et al. The interlayers and effects on injection-production ralationship in fan delta of Badaowan formation, Di-12 Well Area in Cainan Oilfield. Xinjiang Petrol Geol. 2013;34(5):548–551.
Zheng R, Huaguo W, Fengjie Li. High-resolution sequence stratigraphy. Beijing: Geological Publishing House; 2010a.
Zheng Zhan Wu, Changfu SX, et al. Lithofacies and reservoirs of allluvial fan in the Lower Keramay Formation in the block-6 of Karamay oilfield, the Junggar Basin. Oil Gas Geol. 2010b;31(4):463–71.
Zhou L, Yingchang C. The application of clastic grain-size analysis to the genetic study of sand bodies in Upper Es4 Submember of Xin176 Area in Dongying Sag. Aeta Geoscientica Sinic. 2010;31(4):563–73.
Zhu J. The geological cause analysis of high oil saturation in low porosity and low permeability reservoirs—taking Shengli Oilfield in Dongying depression. J Oil Gas Technol. 2008a;30(3):64–7.

Zhu X. Sedimentary petrology (4th edn). Beijing: Petroleum Industry Press; 2008b.
Zhu R, Changmin Z, Fuhua G, et al. Use of sediment dynamic analysis in environment interpretation: A case study on Honghuatao Formation, Upper Cretaceous of Western Jianghan Basin, Hubei Province. Geol J China Univ. 2010;16(3):358–64.
Zhu X, Fen L, Mingxuan T, et al. Diagenesis and genesis of favorable reservoir of fan delta in Eocene Shahejie Formation, Zhanhua Sag Jiyang Depression, Bohai Bay Basin. Geol Rev. 2015;61(4):843–51.

Chapter 4
Geological Genetic Analysis of OGW Systems

When oil, gas, and water have oil–gas, oil–water, and gas–water contacts in a reservoir, they typically behave as a unified oil–gas–water (OGW) system within the deposit. The occurrence and distribution of OGW systems directly impact reserve estimations and development strategies. Therefore, the study of OGW systems is crucial for reservoir characterization (Yinan et al. 1996).

4.1 Research Status in China and Abroad

The research on OGW system runs through the lifecycle of an oil/gas field, from exploration to production. Many scholars have worked on OGW systems (Kewen et al. 1989; Ynadong,1990; Huaqing et al. 1995; Wenxiao et al. 1997; Shangming et al. 1999; Yang 2000; Hong et al. 2000; Jun 2000; Shijia et al. 2001; Yan et al. 2001; Dakang et al. 2002; Tang 2002; Wenbin et al. 2002; Ying 2007; Hu et.al. 2009a, b; Ming et al. 2012; Zhao 2013; Meiyuan 2014; Fucong et al. 2016). Li et al. (1989) calculated the oil–water–gas relative permeability curves according to the pore size probability distribution. Ynadong (1990) used C/O spectral logging method to interpret oil, gas, and water zones and determine the water saturation and water cut of strata in cased wells. Huaqing et al. (1995) investigated the experimental conditions, data processing methods, and analysis techniques for testing oil, gas, and water saturations observed in cores in the laboratory under reservoir pressure and plotted oil and water saturation-degassing correction charts for different oil zones, with an accuracy of 95%. Wenxiao et al. (1997) successfully identified oil, gas, and water zones using acoustic logging data, demonstrating a nonelectric method for evaluating complex formations, such as low-resistivity oil and gas layers and high-resistivity water zones. Futang et al. (1997) elaborated the physicochemical properties, chemical composition, and main ion populations of formation water in northern Songliao

© Petroleum Industry Press 2024

H. Chen, *Geological Genetic Analysis and Application of Oil and Gas Reservoirs*,
https://doi.org/10.1007/978-981-97-7628-3_4

Basin. Shangming et al. (1999) proposed a probability method for composite identification of oil, gas and water zones, which was proved effective when applied in the Puxi and Xinzhan areas, western Songliao Basin. Yang (2000) recognized oil, gas and water zones in reservoirs with low-salinity formation water using the computer technique, based on the (neutron, density, and acoustic) porosity logs and (lateral) resistivity logs, with a coincidence rate of 95% with logging interpretation results. With the Yanqi Basin, Xinjiang, as an example, Hong et al. (2000) build an oil, gas and water zone interpretation model using the back propagation (BP) artificial neural network, enabling the identification of reservoir fluid types based on raw data (e.g. gas logging, geochemistry, and mud logging) and computer-based automatic processing. Jun (2000) used the fuzzy grey correlation analysis to identify oil, gas, and water zones using logging data, with the Tarim Oilfield as an example. Shijia et al. (2001) used the chemical properties of reservoir extracts to identify oil, gas, and water zones, with which the logging interpretation was not affected by the reservoir rock composition and fluid properties. Yan et al. (2001) used the seismic while drilling (SWD) technique to predict the formation pressure and identify the oil, gas, and water zones, and indicated that the SWD technique is advantageous for not interfering the drilling operations, enabling the acquisition of original reservoir data, 24 h monitoring, and providing data about drill bit conditions to reduce drilling risks and about geo-steering drilling of directional wells. Dakang et al. (2002) used the artificial neural network to identify oil, gas, and water zones during mud logging in the Junggar Basin, with the accuracy rate exceeding 90%. Tang (2002) proposed a robust method for prediction of oil and gas in the Carboniferous shallow marine–shore clastic rocks in the Tahe Oilfield. Dakang et al. (2002) conducted correspondence analysis to identify oil, gas, and water zones in the Longxi–Bayan Chagan area, with the coincidence rate of 89.6% to the oil testing results. Ying (2007) performed secondary interpretation and evaluation on the lithology, physical properties, oil-bearing properties and electricity of the Dongying Formation and Shahejie Formation in the Hongxing area of the Liaohe Oilfield and discovered new oil and gas layers that had not been identified in the primary logging interpretation. Yongzhang et al. (2009) distinguished the oil, gas, and water zones by stepwise discriminant analysis and analyzed the oil, gas, water distributions and controlling factors in the Hangjinqi area, the Ordos Basin. Ming et al. (2012) introduced the latest advances in identification of oil, gas, and water zones by the geochemical method, with a case study of the Luxi area in the Junggar Basin. Zhao (2013) proposed a method for identifying oil, gas, and water using the float flow logging map based on the actual liquid production of wells, which has been satisfactorily applied in production interpretation. Meiyuan (2014) applied the well logging curve rebuilding to recognize the oil and gas layers in the Hasshan block logging curves and established an evaluation standard of AC logging curve for oil-bearing properties in shallow clasolites. Fucong et al. (2016) analyzed the lithology, physical properties, oil-bearing properties and electricity of the Heidimiao reservoir in southern Putaohua Oilfield, Daqing, Heilongjiang, based on the core, testing and logging data, and established the oil, gas and water identification chart. Generally, previous studies primarily targeted clastic rock reservoirs and focused on identification of oil, gas, and water zones and description of their distributions. They adopted the methods including geology, geophysics, geochemistry, and production performance, with well logging, mud logging, and geochemical methods in dominance. In the meantime, various mathematical algorithms and computer technologies are increasingly widely used to improve the accuracy in oil, gas, and water zone identification. In the future, more efforts will be made in the detailed logging interpretation of oil, gas, and water by horizon and block, precision geochemical analysis of OGW systems, and 4D subsurface fluid seismic monitoring during exploitation.

Currently, the fine reservoir description mainly includes fine structure interpretation and characterization, and reservoir prediction and description. The attention to the OGW system in reservoirs is far from adequate. Oil and gas are the direct targets in field development, and the characterization of OGW system determines the implementation of various stimulations and the improvement of recovery. Therefore, the OGW system should be considered as a key target in fine reservoir description. In this chapter, the distribution of OGW systems in the Yulou reservoir in a steam flooding test area in the Liaohe Basin's West Sag (hereinafter referred as the Yulou reservoir) is depicted with the support of such methods as the core observation and description, the lithology, physical properties, oil-bearing properties and electricity ("four-indicator") analysis based on core and logging data, the routine analysis of stock tank oil, and the logging interpretation of water-invaded zones. This work provides a basis for shifting from the steam huff-and-puff to the steam flooding in the late period of heavy oil field development. The information on the study area and the target horizons are detailed in Sect. 4.2, Chap. 3.

4.2 Oil-Bearing Properties of Reservoirs

The oil-bearing properties of reservoirs mainly include the oil saturation and occurrence of the reservoirs, which are primarily investigated by core observation and description, logging interpretation, and other methods. Characterization of the oil-bearing properties may facilitate the determination of effective reservoir boundaries, delineation of oil areas, and estimation of resources/reserves.

4.2.1 Description of Oil-Bearing Properties of Core Samples

Cores obtained while drilling often contain oil. Therefore, core observation and description acts as the most direct method to understand the oil-bearing properties of reservoirs. For cores that are confirmed with oil, the oil saturation and occurrence should be described. Fresh surfaces of cores are preferred in the description to accurately determine the oil saturation (Yingchang et al. 2003). The oil-bearing level of the core is mainly determined depending upon the oil occurrence, oil saturation, and oil area. According to Cao et al., there are four oil-bearing levels, including oil sands (oil area > 75%), oil-rich (oil area of 50%–75%), oil-cut (oil area of 25%–50%), and oil patch (oil area < 25%). Based on these criteria, I observed and described the cores (668 m) collected from seven wells in the Yulou reservoir in an area of the Liaohe Basin's West Sag to understand the oil-bearing properties of the reservoirs (Fig. 4.1). The results show that the cores commonly fall at the levels of oil sands and oil-rich, but rarely at the levels of oil-cut and oil patch. This may be attributed to two factors. First, frequent diversions of the underwater distributary channels caused the sand deposits in different periods to overlay one another, and the late channels eroded the early channel sediments, leading to thicker good reservoirs in the target interval. Moreover, the reservoirs are loose, poorly cemented, well sorted, and well rounded. Therefore, they bear oil and gas evenly. Second, cores are mainly collected from the glutenite intervals; hence, some reservoir intervals with poor physical properties may be missed, leading to the less proportion of oil-cut and oil patch cores.

Fig. 4.1 The oil-bearing characteristics of coring wells in Yulou oil layer of a certain area in the western sag of Liaohe Basin. A-J2 well, oil sands, gray-black fine sandstone, depth: 945.39–945.53 m; B-J91 well, oil-bearing, gray-brown gravelly sandstone, depth: 1028.2–1028.25 m; C-J23-261 well, oil-immersed, grayish-brown silty fine sandstone, depth: 962.05–962.25 m; well D-J10-22, oil spot, argillaceous siltstone, depth: 1030.85–1031.05 m

4.2.2 "Four-Indicator" Analysis and Effective Reservoir Delineation

The "four-indicator" analysis is to determine the cutoffs of the effective reservoirs, so as to delineate the oil area in different horizons, thus laying the foundation for reserve estimation. Yang (2014) determined the cutoff of effective reservoir thickness based on four-indicator analysis, with a case study of the Fu-3 Member of the Zhangjiaduo Oilfield, northern Jiangsu Basin. Here, the "four-indicator" relations of the Yulou reservoir are determined through statistical analysis of the description data and log data of 668 m cores from seven wells, and the physical properties of 150 samples. The core samples were taken from yI and yII in seven wells. The results of statistical analysis are representative in both horizontal and vertical oil-bearing properties. It is found that when the porosity and permeability are greater than 28% and 30 mD, respectively, the samples are basically oil-saturated and oil-rich; when the porosity and permeability are less than 28% and 30 mD, respectively, the samples are basically oil-cut and oil-patch. The findings are consistent with the current understanding of the field (Fig. 4.2). Moreover, when the rocks are glutenites, gravel-bearing fine sandstones, or fine sandstones, they are generally oil-saturated; when the rocks are siltstones and mudstones, they are generally oil-cut and oil-patch (Fig. 4.3). Accordingly, it is determined that the porosity and permeability cutoffs of the effective reservoirs in the Yulou reservoir are 28% and 30 mD, respectively.

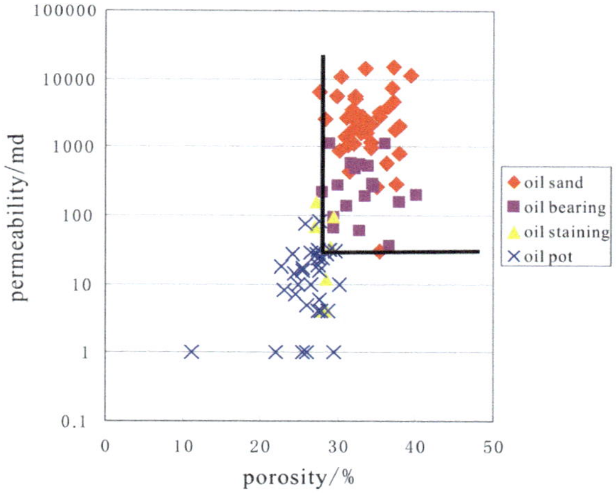

Fig. 4.2 Relationship between oil content and physical properties of Yulou oil bearing sets in some steam flooding experimental area in the West depression of Liaohe Basin

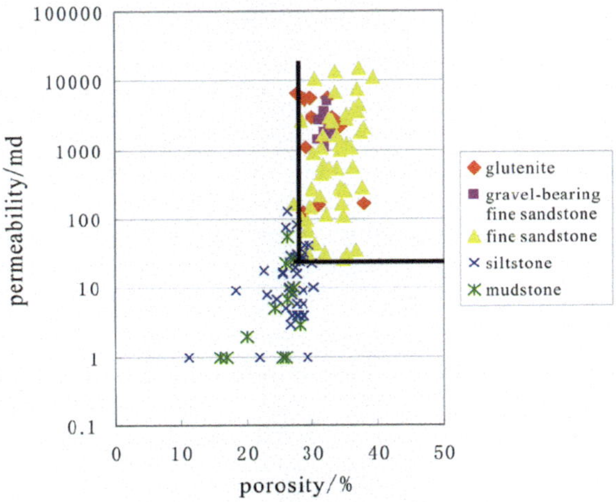

Fig. 4.3 Relationship between the lithology and physical properties of Yulou oil bearing sets in some steam flooding experimental area in the West depression of Liaohe Basin

4.3 Characteristics of Oil Properties

Oil has physical and chemical properties. Physical properties comprise color, density, viscosity, pour point, wax content, solubility, calorific value, fluorescence, optical rotation, etc. Chemical properties include chemical composition, component composition, fraction composition, etc. (Qingquan et al. 2009). Futang et al. (1996) analyze the geochemical characteristics of reservoir fluids and conducted the oil-source correlation for the reservoirs in the Wangfu Sag in the Songliao Basin. Here, the oil properties of the Yulou reservoir are investigated through the analysis of the testing data, and the plane distribution of oil properties is summarized.

4.3.1 Characteristic Parameters of Oil Properties

Heavy oil generally displays high density and high viscosity due to high contents of asphaltene and colloid. These oil properties exert considerable influence on the development of heavy oil reservoirs, so it is essential to analyze the spatial distribution of these properties to efficiently recover heavy oil resources. Analysis of the relationships between different oil property parameters (Fig. 4.4) indicates that, with the increase of the density, the viscosity gradually increases, the contents of colloid and asphaltene also increase, but there is no strong correlation between the wax content and the density. Similarly, as the viscosity increases, the contents of colloid and asphaltene increase. Analysis of the distribution of oil property parameters (Fig. 4.5) shows that the density is rather high (> 0.97 g/cm^3 for most of the 14 samples), the viscosity is mostly greater than 5000 mPa·s, the wax content is rather low (max. 4%,

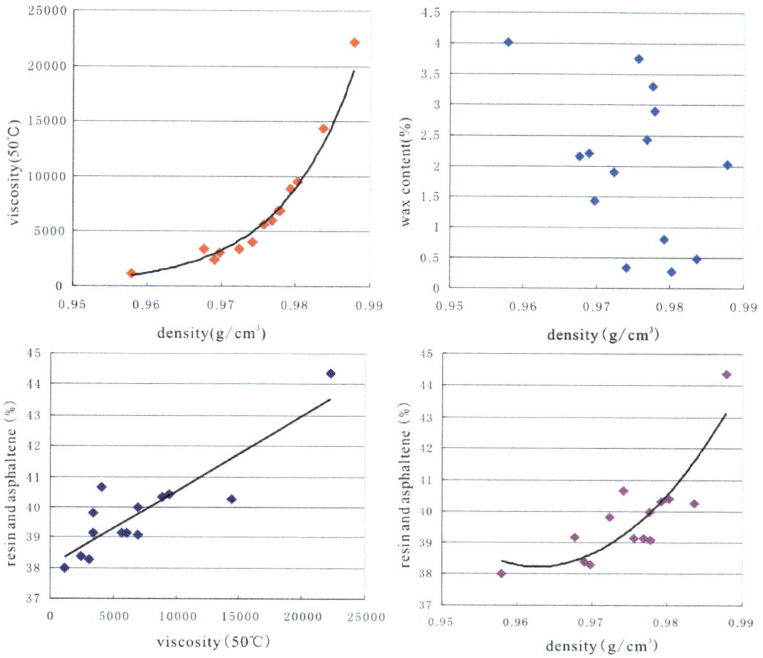

Fig. 4.4 Relationship between crude oil property parameters of Yulou oil bearing sets in some steam flooding experimental area in the West depression of Liaohe Basin

but < 2.5% for most of the samples), and the colloidal and asphaltene contents are high—all above 38%.

4.3.2 *Plane Distribution of Oil Properties*

Analysis shows that the oil density of the Yulou reservoir is rather high—generally > 0.9522 g/cm^3. The oil density in the central part is considerably lower than that in the western and eastern parts of the study area. The oil density tends to increase from northeast to southwest, which is closely related to the NE–SE trend of the monoclinic tectonic setting of the study area. Statistics of the oil viscosity data show that the oil viscosity of the Yulou reservoir exceeds 4000 mPa·s (50 °C) in the central part, but it sharply increases in the southern and eastern parts, with huge viscosity variation observed in a small well spacing locally. Such phenomenon is believed to attribute to multiple factors, such as the formation conditions, temperature and pressure when the reservoir was formed. The wax content distribution is not regular in the middle of the Yulou reservoir, and it is mainly controlled by the nature of the source rocks but not directly associated with other factors. The colloidal and asphaltene contents tend

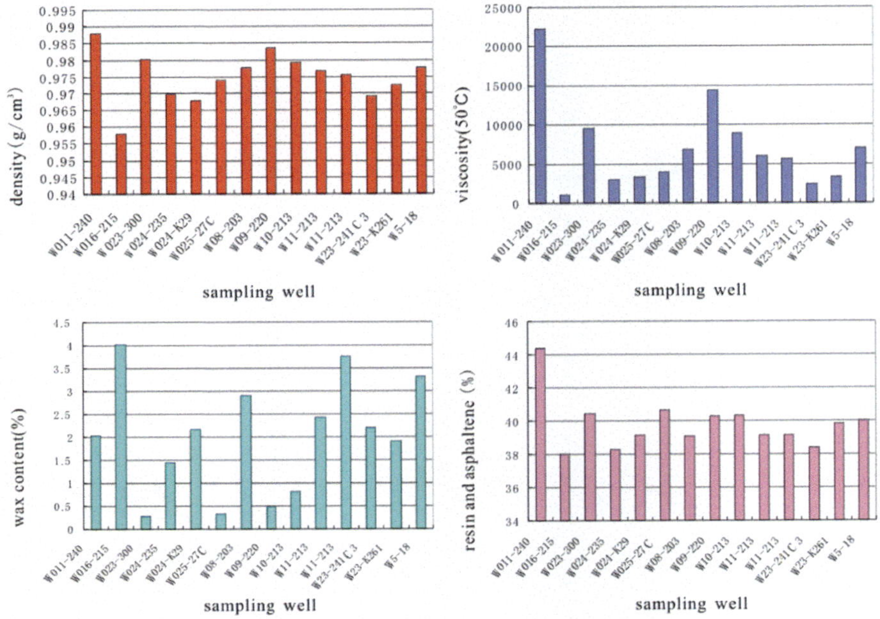

Fig. 4.5 Development rules of crude oil property parameters of Yulou oil bearing sets in some steam flooding experimental area in the West depression of Liaohe Basin

to increase from northwest to southeast, suggesting that they are possibly controlled the distance of oil migration and the reservoir properties, in addition to the nature of the source rocks.

4.4 Characteristics of Formation Water

Formation water refers to the groundwater stored in various forms in the formation voids. In hydrocarbon reservoirs, formation water often exists as edge water, bottom water, or interlayer water (Xuchong 1993; Ye et al. 2009). Formation water has an important impact on the implementation of thermal recovery techniques (e.g. steam huff-and-puff and steam flooding) and the subsurface movement of oil, gas and water during hydrocarbon production. Therefore, the characterization of formation water is a key task in investigating the OGW systems. In this book, the formation water includes edge water and bottom water. Statistical analysis of 27 water samples taken from yI of the Yulou reservoir indicates the remarkable dominance of $Na^+ + K^+$ (the mass concentration of 184–667 mg/L) and HCO^{3-} (the mass concentration of 183.06–1403.5 mg/L) (Table 4.1). The formation water in the study area belongs to the $NaHCO_3$ type, which reflects a semienclosed setting with some free water alternation. The formation water exhibits a low salinity, with a mass concentration of 547.62–2386.2 mg/L.

4.5 Characteristics of Oil–Water Distribution

Oil, gas, and water are always in a relatively stable equilibrium state in connected reservoirs, which have evolved in a long geological history. They are distributed via a gravitational differentiation according to their densities, that is, gas (in oversaturated oil reservoirs), oil, and water are distributed from top to bottom, with gas–oil contact (GOC), oil–water contact (OWC), and gas–water contact (GWC) (Yinan et al. 1996). The spatial distribution of oil, gas, and water can be determined by identifying the locations of GOC, OWC, or GWC, which lay a foundation for reserve estimation and development planning.

4.5.1 Determination of OWC

Determination of OWC is the first task in fine reservoir description for purpose of estimating resources and reserves. Zhengguang et al. (2013) proposed a method for pinpointing the contacts of oil, gas and water by common contour binning stacking, which identifies the contacts without evident seismic reflections by amplifying the amplitude changes generated by fluid change after stacking the three-dimensional seismic waveforms. Here, the fluid contacts are identified by locating the intersections of the pressure gradient lines of different formation fluids through pressure measurement data. According to the repeat formation test (RFT) data (Fig. 4.6), the formation pressure in Wells J013-195 and J10-195 varies greatly at the depth of 1060 m, while the formation pressure in Well J24-310 changes abruptly at 1020 m, confirming the presence of an OWC near these two depths, and the pressure systems above and below the OWC are different. By contrast, the formation pressure changes little in Well J24-K241 at the depth shallower than 1030 m, indicating that the OWC is deeper than 1030 m. Overall, the OWCs within different reservoirs in yI of the study area are located between 1020 and 1060 m. The Yulou reservoir is a lithological-structural reservoir, so the OWC is an interval of value, rather than a fixed value, that is, there are uniform OWCs in the Yulou reservoir.

4.5.2 Sectional Distribution of Oil and Water

A reservoir section of the Yulou reservoir is plotted from the results of the logging reinterpretation (Fig. 4.7). It can be seen that the oil zones are thick and extend widely. The Yulou reservoir presents as a NE–SW monoanticline, where a series of

Table 4.1 Characteristics of water analysis data of Yulou oil layer and Yu I oil layer in a certain area of western sag of Liaohe Basin

	Well Name	Stratum	(Na^+ + Ka^+)	(Mg^{2-})	(Ca^{2-})	(CL^-)	(SO4^{2-})	($CO_3{}^{2-}$)	($HCO_3{}^-$)	mg/L	Total hardness	Total alkalinity	Water property	PH
1	J021-243	y1	363.4	3.65	6.01	141.84	19.21	150	427.14	1111.25	30	600.6	$NaHCO_3$	8
2	J021-243	y1	322	6.08	10.02	88.65	24.02	60	610.2	1120.97	50.1	600.6	$NaHCO_3$	7
3	J026-260	y1	241.5	13.38	42.08	106.38	33.62	90	427.14	954.1	160.2	500.5	$NaHCO_3$	7
4	J26-270	y1	478.4	2.43	12.02	177.3	52.83	150	640.71	1513.69	40	775.7	$NaHCO_3$	8
5	J25-250	y1	310.5	7.3	34.07	88.65	14.41	120	549.18	1124.11	115.1	650.6	$NaHCO_3$	7
6	J022-243	y1	370.3	2.43	16.03	106.38	4.8	60	732.24	1292.18	50.1	700.6	$NaHCO_3$	7
7	J025-23	y1	437	4.86	2	70.92	24.02	0	1037.34	1576.1	25	850.8	$NaHCO_3$	6
8	J24-X32	y1	312.8	3.65	6.01	106.38	9.61	180	305.1	923.55	30	550.5	$NaHCO_3$	7
9	J23-290	y1	239	3.65	8.02	106.38	28.82	60	335.61	781.68	35	375.4	$NaHCO_3$	6
10	J022-263	y1	423.2	3.65	16.03	124.11	24.02	120	701.73	1412.7	55.1	775.7	$NaHCO_3$	7
11	J021-31	y1	653.2	1.22	12.02	124.11	4.8	120	1311.93	2227.28	35	1276	$NaHCO_3$	6
12	J021-31	y1	501.4	7.3	14.03	141.84	28.82	60	1006.83	1760.24	65.1	925.9	$NaHCO_3$	7
13	J021-31	y1	466.9	3.65	18.04	124.11	24.02	150	884.79	1671.51	60.1	875.8	$NaHCO_3$	7
14	J21-270	y1	499.1	3.65	22.04	159.57	28.82	270	549.54	1532.7	70.1	900.8	$NaHCO_3$	6
15	J21-270	y1	370.3	8.51	4.01	106.38	24.02	150	518.67	1183.89	45.1	675.6	$NaHCO_3$	7
16	J20-270	y1	542.8	3.65	12.02	106.38	24.02	240	793.26	1722.13	45.1	1051	$NaHCO_3$	8
17	J24-270	y1	604.9	4.86	14.03	195.03	19.21	180	945.81	1963.84	55.1	1076	$NaHCO_3$	6
18	J24-270	y1	604.9	4.86	14.03	195.03	19.21	180	945.81	1963.84	55.1	1076	$NaHCO_3$	6
19	J22-270	y1	453.1	1.22	12.02	106.38	19.21	210	610.2	1412.13	35	850.8	$NaHCO_3$	8
20	J24-281	y1	271.4	6.08	18.04	106.38	33.62	90	396.63	922.15	70.1	475.4	$NaHCO_3$	7
21	J024-26	y1	395.6	1.22	14.03	124.11	24.02	90	671.22	1320.2	40	700.6	$NaHCO_3$	5

(continued)

Table 4.1 (continued)

	Well Name	Stratum	(Na^+ + Ka^+)	(Mg^{2-})	(Ca^{2-})	(CL^-)	($SO4^{2-}$)	(CO_3^{2-})	(HCO_3^-)	mg/L	Total hardness	Total alkalinity	Water property	PH
22	J24-251	y1	561.2	8.51	22.04	124.11	57.64	60	1189.89	2023.39	90.1	1076	$NaHCO_3$	6
23	J24-251	y1	667	8.51	28.06	106.38	52.83	120	1403.46	2386.24	105.1	1351	$NaHCO_3$	8
24	J27-290	y1	469.2	8.51	10.02	336.87	28.82	90	518.67	1462.09	60.1	575.5	$NaHCO_3$	5
25	J27-321	y1	184	1.22	4.01	70.92	14.41	90	183.06	547.62	15	300.3	$NaHCO_3$	6
26	J25-301	y1	273.7	7.3	2	230.49	52.83	30	244.08	840.4	35	250.2	$NaHCO_3$	6
27	J026-300	y1	243.8	9.73	20.04	53.19	19.21	30	579.69	955.09	90.1	525.5	$NaHCO_3$	7

Note Water analysis time from 2001–2002, the total number of samples of 27

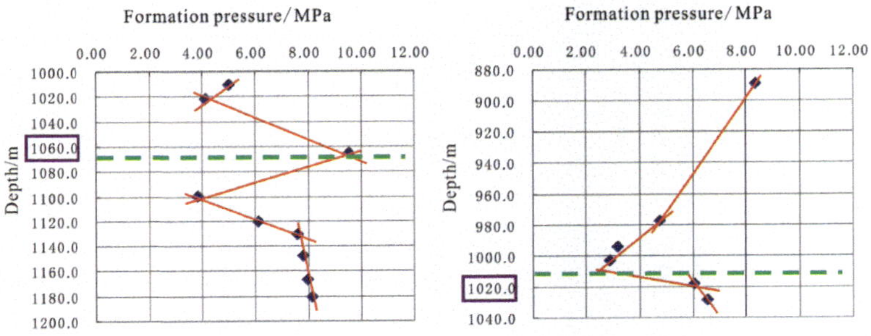

Fig. 4.6 Characteristics of stratum pressure profile of Yulou oil bearing sets in some steam flooding experimental area in the West depression of Liaohe Basin (RFT data, True vertical depth)

oil accumulations are formed under the control of lithology and tectonism. Affected by sedimentation, the sand bodies are well interconnected and the oil zones are continuous in the direction parallel to the provenance; however, the sand bodies are poorly interconnected and the oil zones are discontinuous in the direction perpendicular to the provenance. Comparison of the logging interpretation results with the classification of sedimentary microfacies indicates that the oil zones are mainly deposited in the underwater distributary channel and river-mouth bar settings. Some sediments of underwater distributary interchannel sands have poor physical properties and contain no oil, corresponding to dry zones. The water zones are below the target oil zones, mainly deposited in underwater distributary channel and river-mouth bar settings as well. There is no much difference between the oil reservoirs formed in underwater distributary channels and those formed in river-mouth bars. The positive and negative sedimentary rhythms are the main issue to be investigated during development. The oil zones range in thickness from 1–2 m to 5 m, and the narrow and thin oil zones show much inferior interwell continuity than the thick oil zones.

4.5.3 *Plane Distribution of Oil and Water*

The distribution of oil and water zones is plotted for sublayers based on the fine logging interpretation (Figs. 4.8 and 4.9). It can be seen that the study area has good oil-bearing properties entirely. The oil zones are distributed widely, and thin dry zones are found locally in lumps. In the Yulou reservoir with edge and bottom water, from top to bottom, the oil zones decrease, while the water zones increase. In the lower part of the Yulou reservoir, the water zones become merged, particularly in the east of the study area. The size of the pies in Figs. 4.8 and 4.9 represents the thickness of oil zones. The thickness of oil zone may change remarkably within a short distance. Meanwhile, the oil and dry zones are alternative vertically within one well and the fluid properties change dramatically. These features prove to a certain extent that the delta front reservoirs have strong heterogeneity.

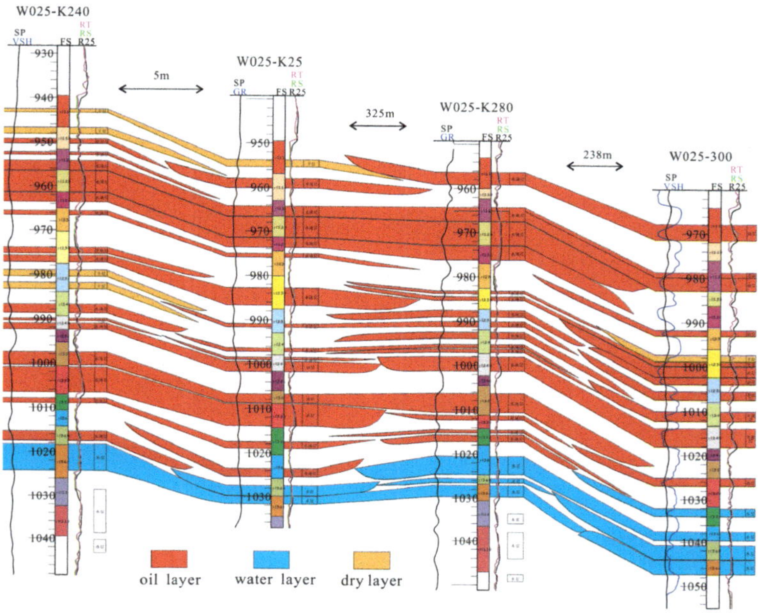

Fig. 4.7 The reservoir section of Yulou oil bearing sets in some steam flooding experimental area in the west depression of Liaohe Basin (Parallel source direction)

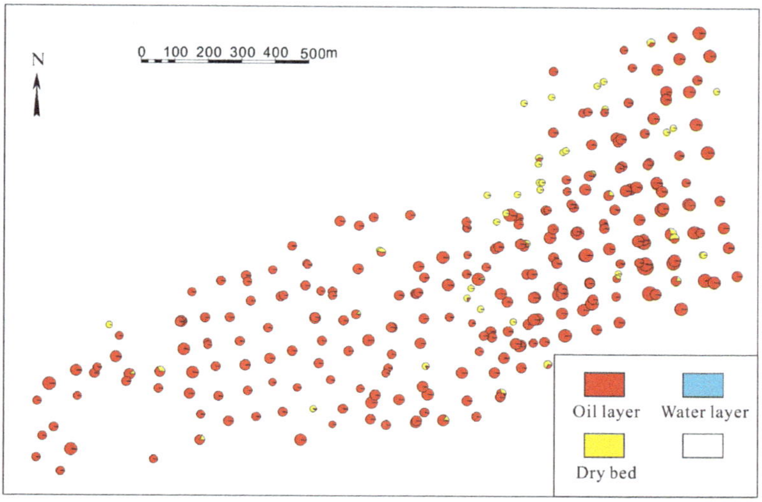

Fig. 4.8 Oil content characteristics of single layer $yI_1{}^{1a}$ of Yulou oil bearing sets in some steam flooding experimental area in the West depression of Liaohe Basin

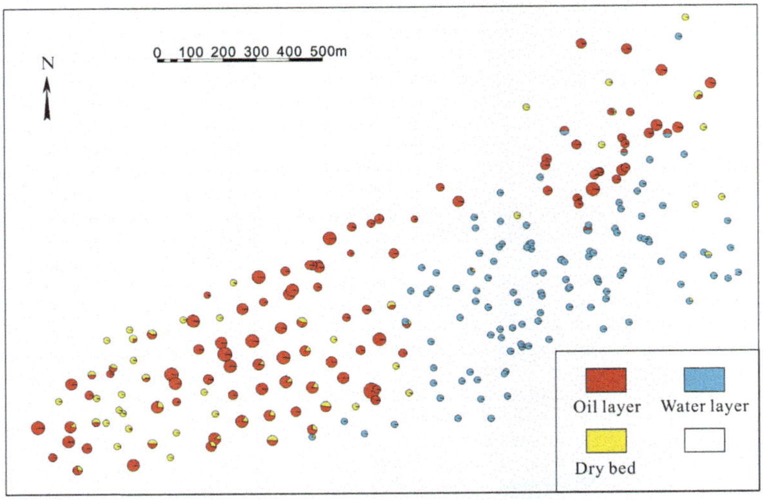

Fig. 4.9 The top structure of a steam drive test area in the West depression of Liaohe Basin in different single layer of the building oil layer

4.6 Impacts of Field Development on OGW Systems and Corresponding Strategies

Influenced by tectonic features and reservoir heterogeneity, the distribution of OGW system is very complex. During the field development, the distribution of the OGW system changes and becomes even more complex after water or steam injection. For instance, the bottom oil zones may be invaded by water. The geological genetic analysis of OGW systems in reservoirs may help to reveal the distribution of OGW system profoundly, providing references for implementation of various development strategies.

4.6.1 Distribution of Water-Invaded Zones

Logging interpretation of water-invaded zone, as an important tool to characterize the remaining oil distribution in the high-water-cut development stage, has always been the focus of researches. It is a technique that can collect the most reservoir information from wellbores, has the widest coverage and the largest sampling density, and is able to reflect various reservoir parameters in real time. It is the primary way to monitor the static and dynamic oil saturations (Peihua 2003). In this section, the distribution of water-invaded zones in the Yulou reservoir is analyzed through fine logging interpretation, and the relevant maps are drawn (Figs. 4.10 and 4.11). The size of the pies in Figs. 4.10 and 4.11 represents the thickness proportion of the oil

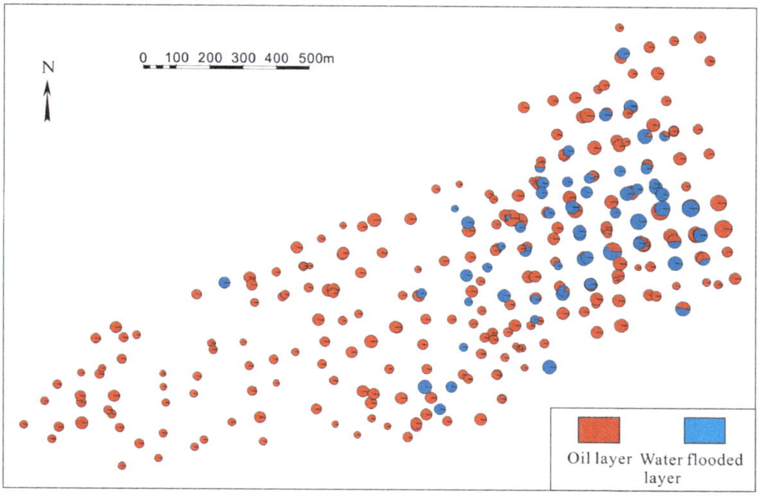

Fig. 4.10 The top structure of a steam drive test area in the West depression of Liaohe Basin in different single layer of the building oil layer

zones or water-invaded zones in the sublayer. It can be seen that the water-invaded zones are concentrated in the right part of the study area, but rare in the left part. The main reason is that the right part is developed earlier than the left part, with denser well pattern and being invaded by edge and bottom water. A steam flooding pilot test is now being carried out in the right part. Similar features are seen in others out of the 29 sublayers in the study area. The closer to the bottom of the reservoir, the more severe the water invasion, which proves to some extent that the water invasion is mainly a result of edge and bottom water channeling. Horizontally, the thickness of water-invaded zones changes greatly within a small well spacing, which is primarily attributed to the strong reservoir heterogeneity affected by the sedimentation and tectonism, and also to the some engineering factors.

4.6.2 Development Strategies Considering the OGW Systems

With the progress of oil and gas development, the reservoir temperature, pressure and other conditions change, and so do the OGW systems. Facing such changes, researchers should fully consider the impacts of the OGW systems in oil and gas reservoirs when designing and implementing various EOR measures. As to the Yulou reservoir, the following strategies should be considered. (1) The perforation horizon and injection pressure are carefully selected/controlled to avoid the bottom water channeling to invade the oil zones in the reservoir, where the water zones are mainly in the lower part, with large thickness and extension. (2) When shifting from steam huff-and-puff to steam flooding or implementing infill drilling, good injector–producer

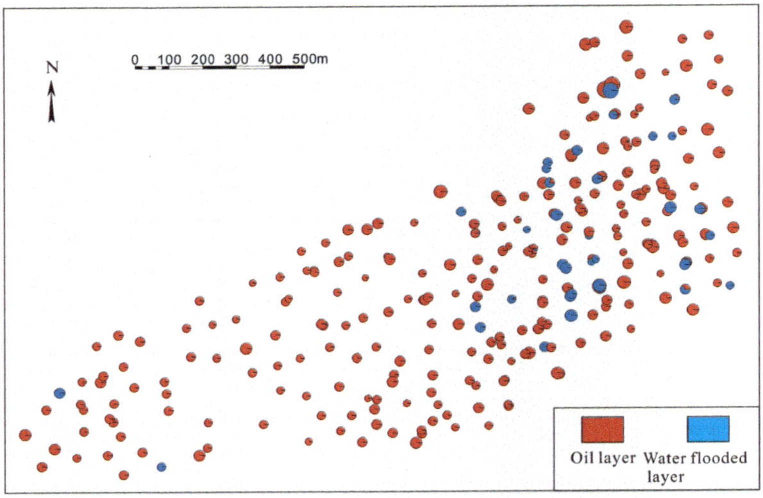

Fig. 4.11 The top structure of a steam drive test area in the West depression of Liaohe Basin in different single layer of the building oil layer

pairs are ensured within one unit depending upon the reservoir profile characteristics and with full consideration to the strong heterogeneity of the fan-delta-front reservoir. (3) During steam injection, the sedimentary rhythm of strata should be considered to integrate the geology and development for purpose of increasing the sweep volume and the oil displacement efficiency, as the oil, gas, and water transport paths are further complicated by steam huff-and-puff or steam flooding. (4) A quantitative database and sedimentary model should be built for the various reservoir architectures of the Yulou reservoir through fine-reservoir architecture investigation in typical pilot blocks and other procedures, providing a basis for infill drilling and adjustment of injector–producer pattern.

The OGW system is a dynamic system influenced by various factors. It is constantly changing with the field development. An integrated analysis combining both static and dynamic information should be done to obtain objective and true knowledge and findings.

4.7 Summary

The Yulou reservoir in the Liaohe Basin's West Sag has good oil-bearing properties – primarily at the levels of oil sands and oil-rich according to the cores. The porosity and permeability cutoffs of the effective reservoirs are 28% and 30 mD, respectively. As the oil density increases, the viscosity increases, and so do the contents of colloid and asphaltene. There is no considerable correlation between the wax content and the oil density. Furthermore, as the oil viscosity increases, the contents of colloid and asphaltene increase. The oil density in the study area is rather large (> 0.97 g/cm^3 in most of the 14 samples); the oil viscosity exceeds 5000 mPa·s; the wax content is rather low, with the max value reaching 4%, but mostly less than 2.5%; the colloidal and asphaltene contents are high, all exceeding 38%.

The formation water in the Yulou reservoir is of $NaHCO_3$ type and alkaline. The OWCs of different reservoirs in yI are between 1020 and 1060 m. Controlled by sedimentation and tectonism, oil zones range in thickness from only 1–2–5 m.

The edge and bottom water in the Yulou reservoir is crucial to field development. The impacts of the OGW systems should be fully considered in the design and implementation of EOR measures. Four strategies should be especially adopted. (1) The perforation horizon and injection pressure are carefully selected/controlled to avoid the bottom water channeling to invade the oil zones in the reservoir. (2) Good injector–producer pairs are ensured within one unit depending upon the reservoir profile characteristics and with full consideration to the strong heterogeneity of the fan-delta-front reservoir. (3) During steam injection, the sedimentary rhythm of strata should be considered to integrate the geology and development for purpose of increasing the sweep volume and the oil displacement efficiency. (4) A quantitative database and sedimentary model should be built for the various reservoir architectures of the Yulou reservoir through fine-reservoir architecture investigation in typical pilot blocks and other procedures, providing a basis for infill drilling and adjustment of injector–producer pattern.

References

Bisong Y. Identification of oil and gas aquifers in low mineralization strata. Nat Gas Ind. 2000;20(2):42–4.

Dakang Z, Guoxi Z. Application of artificial neural network in reservoir identification of logging oil and gas. J Southwest Petrol Inst. 2002;24(3):28–30.

Dongyu Y, Jiangling L, Qingguo X, Jiazheng Z, Xiaohong G. Prediction of formation pressure and oil and gas aquifer with drilling seimic. China Petroleum Explorat. 6(2), 54–56 (2001)

Fucong J, Zhongquan L, Hongkui L, Wei L. Distribution law of oil-gas-water of the Heidimiao reservoir in Putaohua oilfield, Daqing, China. J Chengdu Univ Technol (Sci Technol Edn). 2016;43(6):671–8.

Futang H, Wei T, Zihui, F. Physical and chemical properties and characteristics of formation water in North Songliao Basin. Petrol Geol Oilfield Dev Daqing 1997;16(3):22–5.

Futang H, Zihui F. Oil and gas hydrogeochemical characteristics and oil source comparison in wangfu depression of songliao basin. Petrol Exp Dev. 1996;23(6):28–33.

Hong H, Qiang L, Yuqin X, Cheng Z. BP artificial neural network was used to establish an interpretation model for oil and gas aquifers. Logging Tech. 2000;11(4):13–8.

Huaqing C, Futang H, Ruili N, et al. Study on oil and gas water saturation analysis and degassing correction method of pressure retaining core. Pet Explor Dev. 1995;22(6):73–7.

Jiashan H, Zhiqiang S, Chengzhai L. The cause of gravity anomaly in the south of dongying sag. Petroleum Geol Rec Eff. 2009a;16(2):39–42.

Jun Z. Application of Fuzzy grey correlation analysis method to the identification of oil/gas and water zones. Well Logging Technology. 2000;24(5):337–9.

Jianming T. A method for prediction of oil and gas in carboniferous reservoir, Tahe Oilfield. Geophys. Prospect. Petroleum, 2002;24(3):343–8.

Juntang, Z. Application of float flow log in oil, gas and water identification. J Oil and Gas Technol (J.JPI), 2013;35(10):236–9.

Kewen L, Tongluo Q, Pingping S. Calculation of oil-water-gas relative permeability curves according to pore size probability distribution. Acta Petrolei Sinica. 1989;10(4):65–73.

Meiyuan S. Comprehensive application of well logging curve rebuilding in the recognition of oil and gas layer in the Hassan Block. Sci Technol Eng. 2014;14(19):211–6.

Ming W, Xulong W, Yueqian Z, Baoli X, Wenfang L, Pingan S, et al. Application of identification of oil, gas and water intervals to Luxi Area in Junggar Basin. Xinjiang Petrol Geol. 2012;33(5):543–6.

Peihua Z. Logging technology of water flooded layer in oilfield development. Beijing: Petroleum Industry Press; 2003.

Qingquan Y, Min Y. Interpretation of common terms for oil and gas field development (Third edition). Beijing: Petroleum Industry Press; 2009.

Shangming S, Xiaojie S, Dianjie H. Probability method for comprehensive identification of oil, gas and water layer and its application. Petrol Geol Oilfield Dev Daqing. 1999;18(3):13–5.

Shijia C, Lining M, Feng L. Identifying reservoir fluid types using chemical property of reservoir extracts. WLT. 2001;25(2):136–8.

Wenbin Z, Jingye L, Gaiqin L, Xuerui W. Corresponding analysis of oil and gas aquifer identification methods and applications. Petroleum Geology Oilfield Development of Daqing. 2002;21(6):8–9.

Wenxiao Q, Shuwen Y. Evaluation of the characteristics of pore fluids in permeable formations using acoustic logging data. Well Logging Techn. 1997;21(3):215–20.

Xuchong Y. Hydrogeology of oil and gas fields. Shandong Dongying: Petroleum University Press; 1993.

Min Y. Determination of the four relations and the lower limit of effective thickness in fusan section of zhangjiatao oilfield. Petroleum Geol Eng. 2014;28(2):64–6.

Ying L. Secondary interpretation and evaluation of oil and gas aquifers in hongxing area of liaohe oilfield. Logging Technol. 2007;18(2):29–31.

Yingchang C, Zaixing J. Sedimentology experimental methods and techniques. Beijing : Petroleum Industry Press; 2003.

Yinan Q, Ziqi C. Compiled. Reservoir description. Beijing: Petroleum Industry Press; 1996.

Ynadong T. Fundamentals and application examples of carbon /oxygen spectral logging method for interpretion of hydrocarbon. Geophys Geochem Explorat. 1990;14(5):156–346.

Yongzhang H, Gang L, Yi W, Honghui W, Xinguo D. Distribution law of oil, gas and water, and an analysis of controlling factors of the Hangj inqi area in Ordos Basin, China. J Chengdu Univ Technol (Sci Technol Edn), 2009b;36(2): 28–132.

Zhengguang Z, Ruizhao Y, Yanlong M, Jianxin S, Wei L, Xiaolong W, et al. Method for pinpointing the contacts of oil, gas and water by common contour binning stacking and its application. Nat Gas Geosci. 2013;24(4):808–14.

Chapter 5
Application of Geological Genetic Analysis of Oil and Gas Reservoirs in Field Development

The geological genetic analysis of oil and gas reservoirs is required for various studies related to field development. Its accuracy governs how the genesis of oil and gas reservoirs is precisely understood. This chapter introduces the significance and application of the geological genetic analysis of oil and gas reservoirs in field development, with the case studies of the Ying-1 Member volcanic reservoir in the Xudong area of the Songliao Basin, the glutenite reservoir at the northwestern margin of the Junggar Basin, and the sandstone reservoir in the Liaohe Basin.

5.1 Fine Stratigraphic Division Based on High-Resolution Sequence Stratigraphy

Fine stratigraphic division and correlation is fundamental in the studies related to field development (Xia et al. 1999; Zhang et al. 2011; Chen et al. 2008) and has attached the attention of numerous researchers. Currently, most oil/gas fields in East China are in the mid-late stage of development with high water cut. Therefore, it is necessary to implement separated-layer water/steam flooding and infill drilling to tap the potential of remaining oil and enhance the oil recovery. This underscores the need to build an isochronal stratigraphic framework with high precision. However, the traditional method based on "cycle correlation and hierarchical control" can hardly meet this purpose (Martín-Izard et al. 2009). More advanced techniques/methods are urgently demanded for building a more precise stratigraphic framework. High-resolution sequence stratigraphy is just a typical example. It can divide the strata into layers or even sublayers, so that a high-precision isochronal stratigraphic framework can be built to facilitate the production practices (e.g. remaining oil recovery and EOR) in the mid-late stage of development (Deng et al. 2002; Zheng et al. 2010a, b). Here, the application of high-resolution sequence stratigraphy to fine stratigraphic division in field development is discussed, with a case study of the Yulou reservoir

© Petroleum Industry Press 2024

H. Chen, *Geological Genetic Analysis and Application of Oil and Gas Reservoirs*,
https://doi.org/10.1007/978-981-97-7628-3_5

in a test area in the western slope of the Liaohe Basin's West Sag. The information on the study area and target zone are detailed in Sect. 5.2, Chap. 3.

5.1.1 Sedimentogenesis Guides High-Resolution Sequence Stratigraphic Studies

In China, strata were primarily deposited in continental settings (Qiu et al. 1997), and sedimentary products and features sufficiently recorded by strata indicate the depositional environment. The first step in stratigraphic division and correlation is to seek an initial understanding of strata's spatial distribution, which requires sufficient knowledge of the types of sedimentary facies or even microfacies in the interval of interest. For alluvial fans near the provenance, the sediments deposit rapidly in the piedmont after rushing out of the mountain pass, and the thickness of sedimentary strata varies significantly within a short distance. Later, differential compaction during the diagenesis process further aggravates the thickness variation. This feature should be fully considered in stratigraphic division and correlation. For fluvial deposits in the plain, due to the gentle terrain, the sedimentary facies changes mildly, the stratal deposition is stable, the formation thickness varies only slightly in space, and multiple sets of sand bodies are stacked; hence, the "iso-elevation slicing" method works well in stratigraphic division and correlation. The Yulou reservoir is a fan delta front deposit. Its thickness varies greatly as a result of the frequent diversion of underwater distributary channels during the deposition and the severe truncation of the late channels to the early channels. Moreover, mudstones typically show poor distribution continuity, except for those at the top of yI, the base of yII, and between yI and yII. Additionally, due to the development of tectonic faults, the strata are profoundly broken near the faults, adding the difficulty to stratigraphic division and correlation. Here, based on the classic depositional model of fan delta front (Galloway et al. 1983; Fig. 5.1) and the sufficient knowledge that the formation thickness is spatially highly variable, the high-resolution sequence stratigraphic study is conducted to build a high-precision isochronal stratigraphic framework of the Yulou reservoir.

5.1.2 The Practical Production Needs Govern the Selection of Stratigraphic Division and Correlation Scheme

The first step in high-resolution sequence stratigraphic study is defining the scheme and precision level of stratigraphic division and correlation, which are dependent upon the practical production needs. Essentially, the isochronous stratigraphic division and correlation will ultimately meet the needs of practical production. Field development at various stages requires different levels of precision in the stratigraphic

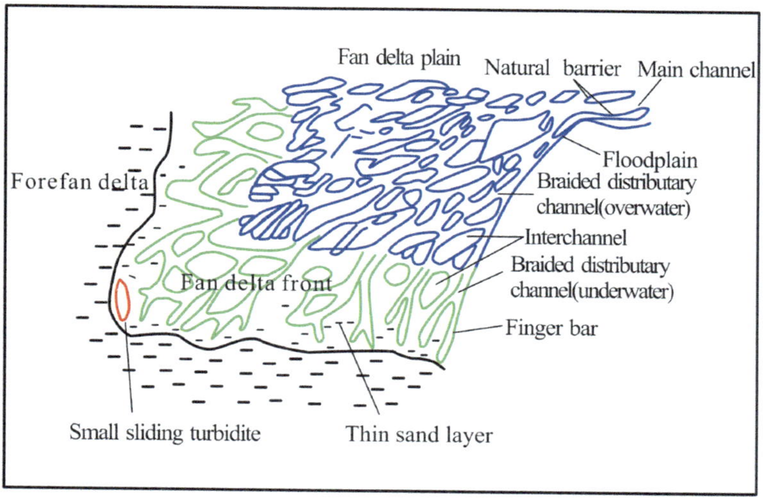

Fig. 5.1 Characteristics of the depositional model of the fan delta (Galloway et al. 1983)

division. At the early stage, the strata are divided at the level of pay zones or layers. At the mid-late stage, especially as the separated-layer water/steam flooding, plugging of the preferred pathways for water/steam channeling and potential tapping of remaining oil are implemented, the strata should be divided at the level of sublayers corresponding to sandbodies. Under this background, the role of high-resolution sequence stratigraphy is becoming prominent. Unlike traditional stratigraphic division and correlation methods, high-resolution sequence stratigraphy theory and method/technique allow delineating the fifth-order sedimentary cycle—the boundary of sublayer—corresponding to the short-term base-level cycle, laying the foundation of fine characterization of sandbodies. For the Yulou reservoir, currently, the production and pressure decline at the late stage of development by steam puff-and-huff. It is urgent to convert the development mode to steam flooding to enhance the heavy oil recovery and tap the potential of remaining oil. Thus, the existing layer-level stratigraphic division scheme should be refined to the sublayer level (corresponding to the fifth-order sedimentary cycle or the fifth-order sequence), aiming to characterize the sandbodies. First of all, the regional marker beds are recognized from seismic data. Then, a large-scale isochronal stratigraphic framework is established after verification and modification based on well–seismic data using calibrated seismic interval velocity and synthetic seismograms. In the context of this framework, the strata are divided after identifying the base-level cycles of different orders on critical wells, based on various well data, including log shape variation, core, sedimentary cycle, and fluid properties. Finally, depending on the skeleton profile network, the abovementioned procedure is extended to the entire study area to enable the fine stratigraphic division and correlation, that is, the interval of interest is subdivided at the level of sublayer, so that a high-precision isochronous sequence stratigraphic framework is built for the study area.

The term "sublayer" refers to an independent sand zone storing oil/gas, separated from the overlying and underlying strata by barriers, within which there is an independent fluid flow unit (Qiu 1997). A critical point in the high-resolution sequence stratigraphic study is choosing an acceptable stratigraphic division scheme. This book presents a stratigraphic division to the level of sublayer. It is believed that the sublayer division scheme be determined according to eight principles: (1) at least 50% of the sublayers correspond to one sand body; (2) the sand thickness in a sublayer is not greater than the maximum thickness of channel sands of a single phase, in order to ensure the ability to separate the vertically stacked sand bodies from one another; (3) the well pattern is sufficiently dense at a scale equivalent to or less than a single channel, in order to separate channels of different phases; (4) the boundary of a sublayer usually corresponds to the maximum value on the conductivity log and indicates the location of lake flooding surface; (5) the geological age of a sublayer is approximately 0.03 Ma; (6) the sublayers roughly correspond to the mid- to short-term base-level cycle in high-resolution sequence stratigraphy; (7) the sublayers basically correspond to the fifth-order sequence outlined by Vail in his classic sequence stratigraphy; and (8) the response model of short-term base-level cycles at critical wells can be a reference to delineate the boundary of most sublayers in uncored wells (in other words, it is possible to delineate the short-term base-level cycle in uncored wells). Notably, determining the scheme of sublayer division is not easy. Instead, an initial scheme should be designed depending upon various information, such as log shape and sedimentary cycle; then, the scheme is extended from critical wells to other wells on the skeleton profile network, and then to the entire skeleton profile network after being revised if necessary; the procedure is repeated until the final scheme is determined.

The most applied classification system with three orders of base-level cycles (long-term, mid-term, and short-term) is adopted in this book according to the practical production needs. The Yulou reservoir in the Paleogene Sha 1 Member was deposited at approximately 1.2 Ma (Zheng et al. 1999). In this book, it is divided into two long-term base-level cycles (yI and yII), which can be subdivided into 12 mid-term base-level cycles (yI_1^{1}, yI_1^{2}, yI_2^{3}, yI_2^{4}, yI_3^{5}, yI_3^{6}, yII_1^{1}, yII_1^{2}, yII_2^{3}, yII_2^{4}, yII_3^{5}, and yII_3^{6}). These mid-term base-level cycles are further subdivided into 29 sublayers (yI_1^{1a}, yI_1^{1b}, yI_1^{2a}, yI_1^{2b}, yI_1^{2c}, yI_2^{3a}, yI_2^{3b}, yI_2^{3c}, yI_2^{4a}, yI_2^{4b}, yI_2^{4c}, yI_3^{5a}, yI_3^{5b}, yI_3^{5c}, yI_3^{6a}, yI_3^{6b}, yI_3^{6c}, yII_1^{1a}, yII_1^{1b}, yII_1^{2a}, yII_1^{2b}, yII_2^{3a}, yII_2^{3b}, yII_2^{4a}, yII_2^{4b}, yII_3^{5a}, yII_3^{5b}, yII_3^{6a}, and yII_3^{6b}) corresponding to short-term base-level cycles.

5.1.3 Integrating Well and Seismic Data Forms the Basis for High-Resolution Sequence Stratigraphic Studies

Well data and seismic data are the most fundamental information for stratigraphic division and correlation. Well data provide detailed information on a vertical succession near a wellbore, and seismic data characterize the macroscopic development of strata. Exploration-related studies emphasize seismic data because well data are typically minimal. As petroleum exploration and development studies progress, the availability of more well data and the requirement for a higher level of precision highlight the importance of well data. These data must be combined in a high-resolution sequence stratigraphic study to build a high-precision isochronal stratigraphic framework applicable to the mid- to late-stage development.

Tracking and recognizing the regional marker bed on the seismic section is the first step to avoiding the diachroneity associated with fine stratigraphic division and correlation in field development. The regional marker bed is calibrated with the well using formation velocity or synthetic seismogram calibration. Thus, integrating well and seismic data provides constraints on well-based stratigraphic division and correlation. For the Yulou reservoir, in this book, the regional marker bed is first recognized with seismic data (Fig. 5.2). Three sets of regional marker beds are identified: (1) the top of the Yulou reservoir, (2) the bottom of the Yulou reservoir, and (3) the boundary between yI and yII. The top of the Yulou reservoir presents a set of mudstone, which has a distinct boundary with the overlying basalt at the bottom of the Dongying Formation and exhibits strong reflection on the seismic section. Both the bottom of the Yulou reservoir and the boundary between yI and yII are mudstones which are stable regionally and show strong seismic reflection, but weaker than the mudstone at the top of the Yulou reservoir. The regional marker beds are readily recognizable on well logs. The top of the Yulou reservoir (or the bottom of the Dongying Formation) is interpreted to be a set of dark gray basalt, with low induction and acoustic values, and presenting a straight spontaneous potential (SP) curve. The boundary between yI and yII is interpreted to be a set of bell-shaped gray mudstone, with low induction and high acoustic values, and presenting straight SP and bell-shaped induction and resistivity curves. The bottom of the Yulou reservoir is interpreted to be a set of funnel-shaped mudstone, staying as the boundary between the Yulou reservoir and the Xinglongtai reservoir, with extremely low induction and high acoustic values and presenting a funnel-shaped induction curve and a straight SP curve. Seismic interval velocity calibration extends the seismic-recognized marker beds to wells to verify and correct them with well logs. Accordingly, a large-scale isochronal stratigraphic framework is built with consistency between well and seismic, permitting the subdivision of sublayers in the Yulou reservoir. Inconsistency may exist between well-based and seismic-based sequence delineation results because of their different precision levels and sequence delineation bases. Seismic stratigraphy relies heavily on recognizing unconformity represented by discordance. Well-based sequence delineation is limited by the fossil zone's accuracy and based on the combination of lithology and resistivity. This practice will frequently fail to pick local unconformities because there is

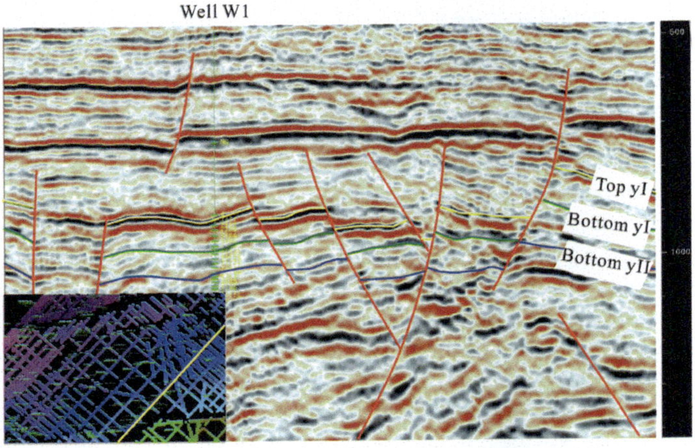

Fig. 5.2 Establishment of large-scale sequence stratigraphic framework of Yulou oil layer in a certain area of western sag in Liaohe Basin by well-seismic combination

no significant difference in lithology and resistivity of the sequences above and below. As a rule, the properties of seismic and well-delineated sequences are inconsistent. The former is the chronostratigraphic unit, and the latter is the lithologic-stratigraphic unit. Seismic data shall prevail when irresolvable conflict occurs between them (Chen et al. 2009).

5.1.4 Building the Response Model of Base-Level Cycles for Critical Wells Is Crucial in High-Resolution Sequence Stratigraphic Studies

Building a response model of base-level cycles for critical wells is crucial in high-resolution sequence stratigraphic studies (Fig. 5.3). Sublayers cannot be divided using seismic data due to resolution. High-resolution sequence stratigraphic study recognizes and delineates various orders of base-level cycles. Multiple classification schemes of base-level cycles have been reported (Deng et al. 2002; Zheng et al. 2010a, b). In the study area, there are approximately 400 wells, where the transfer point of short-term base-level cycle, that is, the location of the lake flooding surface, is delineated mainly according to the conductivity curve. A base level is a continuous potential energy surface relative to the surface relief rather than a physical interface. It reflects the degree of disequilibrium between the Earth's surface and the surface processes seeking equilibrium. A genetic sequence is a stratigraphic unit comprising aggradational/progradational deposits accumulated during a complete base-level cycle, characterized as the change from increasing to decreasing accommodation space. The changes in the base level (rise to fall or fall to rise) mark the semicycle boundary of a genetic sequence. Regardless of scale, stratigraphic cycles created by base-level cycles are time-stratigraphic units, because they are constituted by stratigraphic records accumulated in genetically linked depositional environments during the base-level cycle change. A base level below the Earth's surface initiates erosion, so the duration of the base-level cycle runs the gamut from stratigraphic (rock) records to and hiatus in sedimentation. Thus, the recognition and delineation of varying orders of base-level cycles provide the basis for constructing a high-resolution stratigraphic framework (Deng et al. 2002).

Lake flooding surfaces can be identified on conductivity logs, as discussed in this book. The gamma-ray (GR) log, typically sensitive to the sand content in the rock and can distinguish between sandstone and mudstone and characterize the cyclicity induced by sand content change, is not applicable to the Liaohe Basin's West Sag because the sediments from the source contain radioactive components. In contrast, the induction log is one of the most used logging sequences (Deng et al. 2002). In practice, varying orders of lake flooding surfaces (especially the transfer points of short-term base levels, that is, the fifth-order sequence boundary) are first identified based on the conductivity log. Meanwhile, the response models of mid-term and short-term base-level cycles for critical wells are built according to the spontaneous potential (SP), resistivity (RT), acoustic (AC), and density (DEN) logs (and the vertical combinations of cycles and the lithologic contrasts presented by cores, in case of cored well). With these models, the base-level cycles (especially the short-term base-level cycles) are delineated. Then, the practice is extended to the entire study area according to the skeleton profile network, enabling the sublayer division and correlation. When short-term cycles are delineated on critical wells, the vertical combinations of cycles presented by cores should be considered (Fig. 5.3). This consideration provides greater precision in recognizing base-level cycles while eliminating the systematic errors associated with well logs, such as instrument and human errors. Building a response model of base-level cycles for critical wells is crucial for applying the high-resolution sequence stratigraphy theories and methodologies to the stratigraphic division in field development.

5.1.5 The Traditional Stratigraphic Division and Correlation Method Supplements High-Resolution Sequence Stratigraphy

It is noted that high-resolution stratigraphy is not incompatible with the traditional method based on cycle correlation and hierarchical control. High-resolution stratigraphy is the extension and upgrading of the traditional method, and it shows more implications of the origin of strata and makes the stratigraphic division and correlation more scientific and credible. Integrating, corroborating, and supplementing the high-resolution sequence stratigraphy with the procedures of the traditional method,

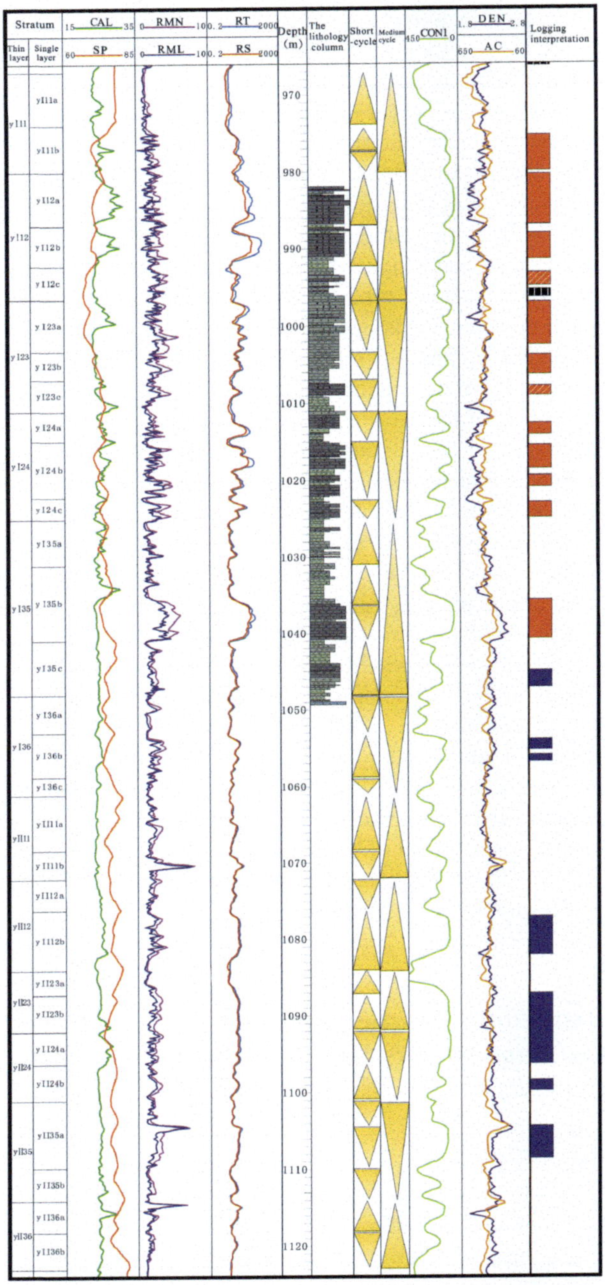

Fig. 5.3 Single well short-term base-level cycle response model of W2 well in Yulou oil layer in a certain area of western sag in Liaohe Basin

such as marker bed correlation, sedimentary cycle characteristics, log shape variation, and reservoir fluid property variation, can improve the fine stratigraphic division and correlation (Fig. 5.4). For instance, widespread mudstones at the top and bottom of the Yulou reservoir and between yI and yII are regional marker beds. They ensure the accuracy of the large-scale isochronal stratigraphic framework and form a solid foundation for further fine stratigraphic division and correlation based on high-resolution sequence stratigraphy. Also, sublayer division for uncored wells benefits from fluid property variations. Fine stratigraphic division and correlation in the development stage is a multi-information job, and it requires the close integration of high-resolution stratigraphy with the traditional methods/techniques.

With the five aspects above considered, high-resolution sequence stratigraphy can be effectively applied to fine stratigraphic division and correlation in field development. In this book, the thickness map is prepared for each sublayer based on the stratigraphic division and correlation results (Fig. 5.5). As shown in Fig. 5.5, $yI_1{}^{1a}$ and $yI_1{}^{1b}$ exhibit the thickness distribution roughly in strips, trending NW–SE, which is consistent with the principal orientation (NW) of the provenance of the Yulou reservoir. Local high-value zones are in lumps, mainly resulted from the frequent diversion of underwater distributary channels during sedimentation and the differential compaction during late-stage diagenesis. The rest 27 sublayers are observed with

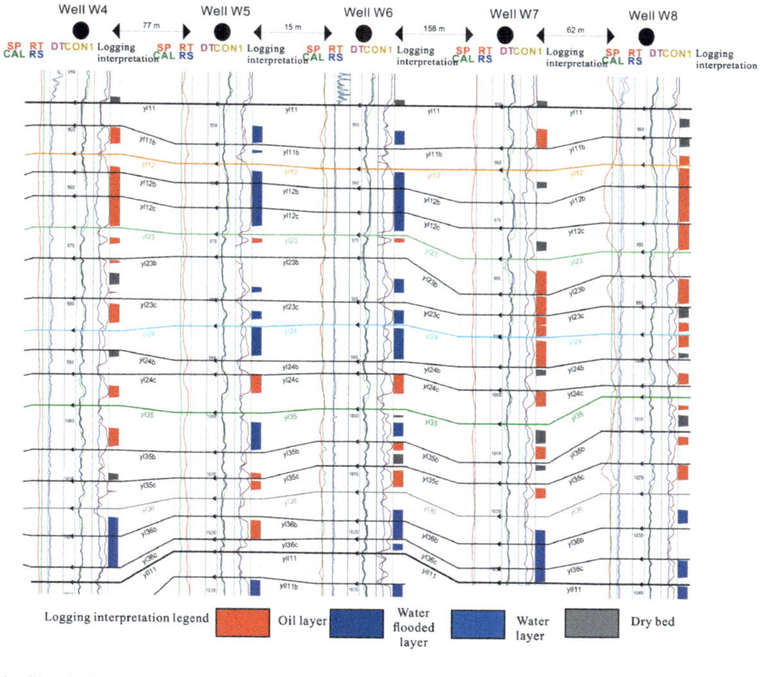

Fig. 5.4 Single-layer high-resolution sequence stratigraphic analysis results of Yulou oil layer in a certain area of western sag in Liaohe Basin

similar features of thickness distribution, despite the variation at different locations. This finding justifies the fine stratigraphic division and correlation in this book.

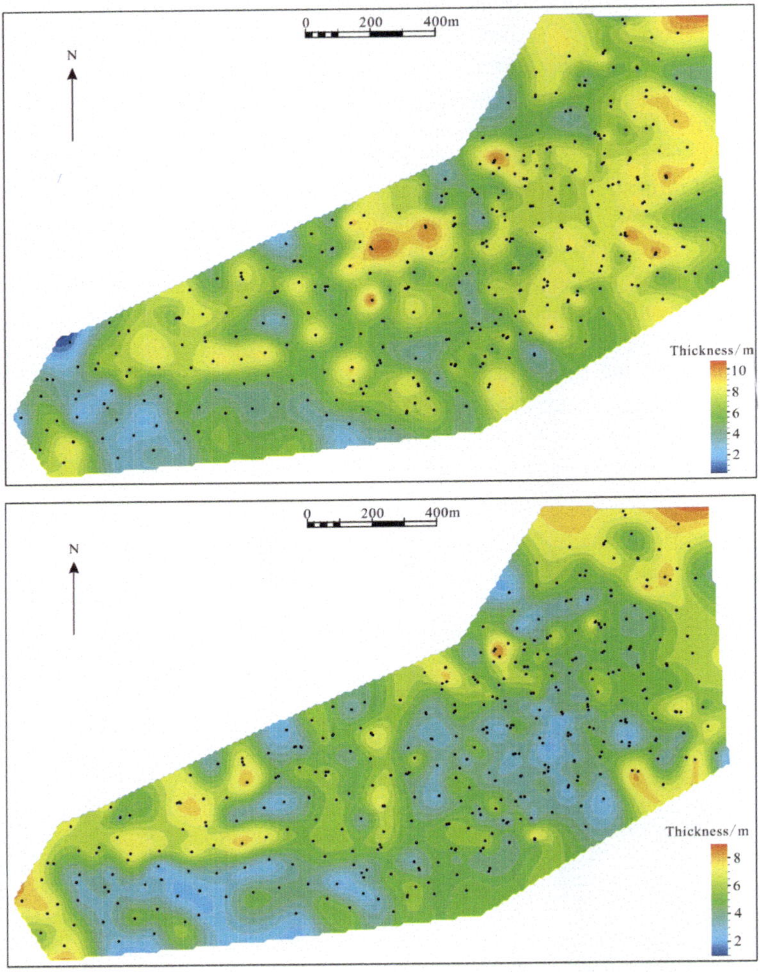

Fig. 5.5 The plane distribution characteristics of the thickness of two single-layer strata in the yI11 layer of Yulou oil layer in a certain area of the western depression of Liaohe Basin. A-single layer yI11**a** formation thickness plane distribution map; **b** single layer yI11b formation thickness plane distribution diagram

5.2 Control and Influence of Sedimentary Genetic Types on Glutenite Reservoir Architectures

The reservoir architectures can effectively reflect the heterogeneity, connectivity, and other attributes of reservoirs, and thus have strong controls on field development (Chen et al. 2015a, b). Therefore, architecture characterization has become a critical subject in researches relating to field development. Investigation of sedimentary facies, especially the analysis of sedimentary genesis, is the basis of reservoir architecture characterization. Alluvial fan is a fan-shaped accumulation of coarse-grained clastics extending from mountain streams into the piedmont basins (Galloway et al. 1983; Mei 1994). Spatially, it is a huge conical deposit expanding along the mountain pass, with the top of the cone pointing to the mountain pass and the bottom facing the plain, which can reach hundreds to hundreds of kilometers. Alluvial fans are concave-downward lens- or wedge-shaped on vertical section, and convex on horizontal section, with the surface slope up to 5°–10° at the fan head near the mountain pass and of 2°–6° farther away from the mountain pass (Zhu 2008a, b). Alluvial fans are widely developed in the Permian and Triassic of the Junggar Basin, the Cretaceous of the Jiuquan Basin, and the Paleogene and Neogene of the Bohai Bay Basin in China (Jiang 2003a, b). The Karamay Oilfield at the northwestern margin of the Junggar Basin, adjacent to the Zaire Mountains to the west, is a monoclinic structure striking NE–SW, approximately 50 km long and 10 km wide. From a terrain perspective, it descends step-like from northwest to southeast. The faults are well developed and divide the oilfield into nine areas and several fault blocks. The No.6 area is the study area in this book. The Triassic Lower and Upper Karamay Formations are the major oil-bearing systems. The Lower Karamay Formation is divided into two pay zones: S6 and S7, of which nine layers ($S6^3$, $S7^1$, $S7^{2-1}$, $S7^{2-2}$, $S7^{2-3}$, $S7^{3-1}$, $S7^{3-2}$, $S7^{3-3}$ and $S7^4$) are the targets here. The Lower Karamay Formation is 350–850 m deep with a thickness of 50–70 m. The study area has 1085 oil producers and water injectors averagely spaced at approximately 110 m (Zheng et al. 2010a, b). Previous studies indicate that the alluvial fans at the northwestern margin of the Junggar Basin were deposited in arid setting (Jiang 2003a, b; Pang et al. 2004; Bao et al. 2005; Liu et al. 2006a, b; Huang et al. 2007; Zhu 2008a, b; Wu et al. 2008; Sosng et al. 2008; Zheng et al. 2010a, b; Gong et al. 2010, 2011; Li et al. 2010a, b; Qin et al. 2010; Gao 2011). However, the analysis of petrology, sedimentary components, sedimentary structures, textural maturity, sedimentary cycles and fan scale suggests that the alluvial fans of the Lower Karamay Formation in the No.6 area at the northwestern margin of the Junggar Basin lie between wet fans and arid fans, as a transitional type. The information on the study area and the target horizons are detailed in Sect. 5.2, Chapter 3.

5.2.1 *Classification and Characteristics of Alluvial Fans*

Two types of alluvial fans have been recognized, namely, arid fans and wet fans (Galloway et al. 1983; Mei 1994; Jiang 2003a, b; Zhu 2008a, b). Mei (1994) reported that common wet fans are crossed by perennial rivers and covered by adequate vegetation, but the extent of wet fan depends on the severity of periodic flooding, allowing its larger area than arid fan and small surface slope. Debris flow exerts weaker effect on the alluvial fan, but influences just the fan head. The conglomerate and sandstone are high mature, and the coal and seat earth are relatively developed. The rock fragments become finer downstream. The gravels are concentrated in the fan head, primarily in clast-supported massive formations, and they are well-rounded. The mid-fan sediments comprise interbeds of sandstone, glutenite, and conglomerate, in which the sandstone is primarily of braided channel facies, with trough and tabular cross beddings, as well as parallel bedding and ripple cross bedding locally. Unlike wet fans, arid fans are small-scale conical accumulations that are convex horizontally and concave vertically. The sediments in arid fans are primarily red. No internal structure is observed in the fan head, but only the imbricated gravels can be observed locally. The debris-flow deposits are widely distributed laterally, and the trough-flood deposits have eroded surfaces at the bottom, showing good sorting. Scour-filling structures are developed in the mid-fan, and sandy sediments are dominant in the lower fan.

5.2.2 *Types of Alluvial Fans in the Lower Karamay Formation at the Junggar Basin's Northwestern Margin*

5.2.2.1 Petrology

The cores were recovered from 11 sealed-coring wells in the study area. Figure 5.6a, b and c show the mudstone cores from Well J6, and Fig. 5.6d shows the mudstone core from Well J5. It can be seen that the cores in $S7^4$–$S7^1$ are gray-black, gray-green, or gray in color, and only a few cores are red due to drilling fluid contamination (not thoroughly washed) on surfaces, indicating that the mudstones were deposited in a reducing environment rather than an oxidizing environment. In $S6^3$ of some wells, the mudstones are reddish brown, indicating an oxidizing environment. Therefore, it is preliminarily determined that the depositional environment in $S7^4$–$S6^3$ gradually changes from a reducing environment to an oxidizing environment, and correspondingly, the alluvial fan evolves from a wet fan into an arid fan.

Fig. 5.6 Core characteristics of Kexia Formation reservoir in a certain area of northwestern margin of Junggar Basin **a** brownish red mudstone, 392.0–392.15 m; **b** grey mudstone (red for drilling fluid), 401.43–401.78 m; **c** grey mudstone (red for drilling fluid), 392.18–392.48 m; **d** dark gray black mudstone, 395.93–396.28 m

5.2.2.2 Sedimentary Components

Coal seams and seat earths are not seen in all cores of $S7^4$–$S6^3$, indicating that the alluvial fans of the target layers are not typical wet fans. However, the mudstones of $S7^4$–$S7^1$ maintain reducing colors, such as gray or gray-green, indicating that the alluvial fans from S7 to S6 are not typical arid fans but transitional between wet fans and arid fans.

5.2.2.3 Sedimentary Structur*es*

Figure 5.7a, b show the sedimentary structures of the cores from Wells J1 and J7, respectively. The scour-filling structures are observed in the alluvial fans of the target layers, reflecting the characteristics of arid fans (rather than wet fans). The findings indicate that the target layers do not belong to typical wet fans. Figure 5.8a, b show the sedimentary structures of the cores from Wells J7 and J3, respectively. It can be seen that typical sedimentary structures, such as tabular cross bedding and trough cross bedding, are developed, reflecting the characteristics of typical wet fans. The findings indicate that the alluvial fans of the target layers are not typical arid fans. It is thus inferred that the alluvial fans of the target layers are transitional types between arid fans and wet fans, with some characteristics of both types (Figs. 5.7 and 5.8).

Fig. 5.7 Characteristics of core erosion-filling structure of Kexia Formation in a certain area of northwestern margin of Junggar Basin A scouring surface, 405.81–406.02 m; B scouring surface, 408.39–408.89 m

Fig. 5.8 Characteristics of core bedding structure of Kexia Formation in a certain area of northwestern margin of Junggar Basin A-grooved cross-bedding, 418.92–418.98 m; B- tabular cross-bedding, 420.05–420.2 m

Fig. 5.9 Characteristics of structural maturity of Kexia Formation sediments in a certain area of northwestern margin of Junggar Basin A - dark brown medium conglomerate, 398.61–398.80 m; B- light brown sandy conglomerate, 402.40–402.51 m

5.2.2.4 Textural Maturity

The cores from Well J1 reveal that most of the conglomerates in the target layers are poorly sorted and show low textural maturity (Fig. 5.9), different from the feature of wet alluvial fans (with high textural maturity). Thus, it is inferred that the alluvia fan of the target layer is not a typical wet alluvial fan.

5.2.2.5 Sedimentary Cycles

As to sedimentary cycles of the target layers in the study area, $S7^4$–$S6^3$ present as a retrogradational upward-fining cycle, indicating that the alluvial fan of the target layers is a retrogradational succession. This suggests that the river gradually shrinks as the climate changes from humid to arid, with reduced sediment supply and downsizing alluvial fan. At the same position on the plane, coarse-grained sediments, such as channel flow gravel and sheetflood inner sand, are developed in $S7^4$ in the lowermost part of the Lower Karamay Formation. In $S7^{3-3}$ and $S7^{3-2}$, coarse-grained sediments develop, such as laminar flow gravel and sheetflood outer sand. Upwardly, sediments such as braided channel, braided sand/gravel bar, and cross flow sand occur with finer grain sizes than in $S7^4$. In $S6^3$, interchannel fine grain is dominant, and thin-layered runoff channel occurs, with the grain size finer than the fourth-order architectures such as braided channel and braided sand/gravel bar, indicating that the Lower Karamay Formation gradually transits from coarse-grained fourth-order architectures to fine-grained fourth-order architectures. The changes in the plane and profile are consistent. The $S7^4$ deposition occurred in the fan head, the $S7^2$ deposition in the mid-fan, while the $S6^3$ deposition in the lower fan, indicating the gradual regression of the alluvial fan. This proves the transition of the alluvial fan depositional setting from humid to arid.

5.2.2.6 Fan Scale

The alluvial fan of the target layers can reach 99 km^2 in scale. The arid alluvial fan is usually a cone with a small area (<100 km^2), in contrast to the larger area of the wet alluvial fan (Li et al. 1997; Zhu 2008a, b). Generally, the alluvia fans of the target layers show the scale at the boundary between the arid fans and the wet fans, indicating possibly a transitional type between the two. Judging from the sediment characteristics, the alluvial fans of the target layers are intermediate between wet fans and arid fans, characterized by small scale, with sandy deposits absent in the lower fan, and less runoff channels which are small in scale (< 5 m thick and 80–130 m wide), and dominated by fine-grained interchannel muddy or silty sediments (Fig. 5.10). Overall, the alluvial fans of the target layers are believed to be transitional between wet fans and arid fans, primarily from the late depositional stage of the wet fan (the S7 depositional stage) to the initial depositional stage of the arid fan (the $S6^3$ depositional stage).

5.2.3 Influence of Alluvial Fan Types on Oil and Gas Development

The alluvial fan type is a scientific issue and significantly affects production practices. Regarding sedimentary origin, the alluvial fan of S7 was controlled by fluvial process, with the debris flow (mud flow) restricted in the fan head; the alluvial fan of $S6^3$ contains debris flow (mud flow) sediments, but it is dominated by runoff channels (interchannel sediments, traction flow rather than gravity flow), since the deposition occurred in the lower fan. The type of alluvial fan plays a critical role in controlling the morphology of sedimentary sand bodies (Mei 1994). The wet alluvial fan is more affected by fluvial process, so it is sheet-like; the arid alluvial fan is dominated by debris flow, so it is mostly thick and wedge-shaped (Fig. 5.11, Mei 1994). Sandbody morphology is crucial to the research on oil and gas development in the conglomerate reservoirs of the target layers. In terms of architecture analysis, for example, the sandbody's shape is of great significance for the accurate prediction of sandbodies, the analysis of sandbody connectivity between wells, and the characterization of spatial architectures of glutenites. For the wet fan, sheet-like sandbodies formed by the fluvial process are better connected spatially than other sandbodies, and may easily form dominant pathways during water injection, due to the main stream line (Jin 2009), via which the injected water may channel. For the arid fan, the wedge-shaped sandbodies formed by debris flow show poor physical properties and worse spatial connectivity than other sandbodies; hence, the probability of injected water channeling along the dominant pathways is slight. Therefore, different development measures should be taken depending upon the sedimentary sandbodies formed by different types of alluvial fans.

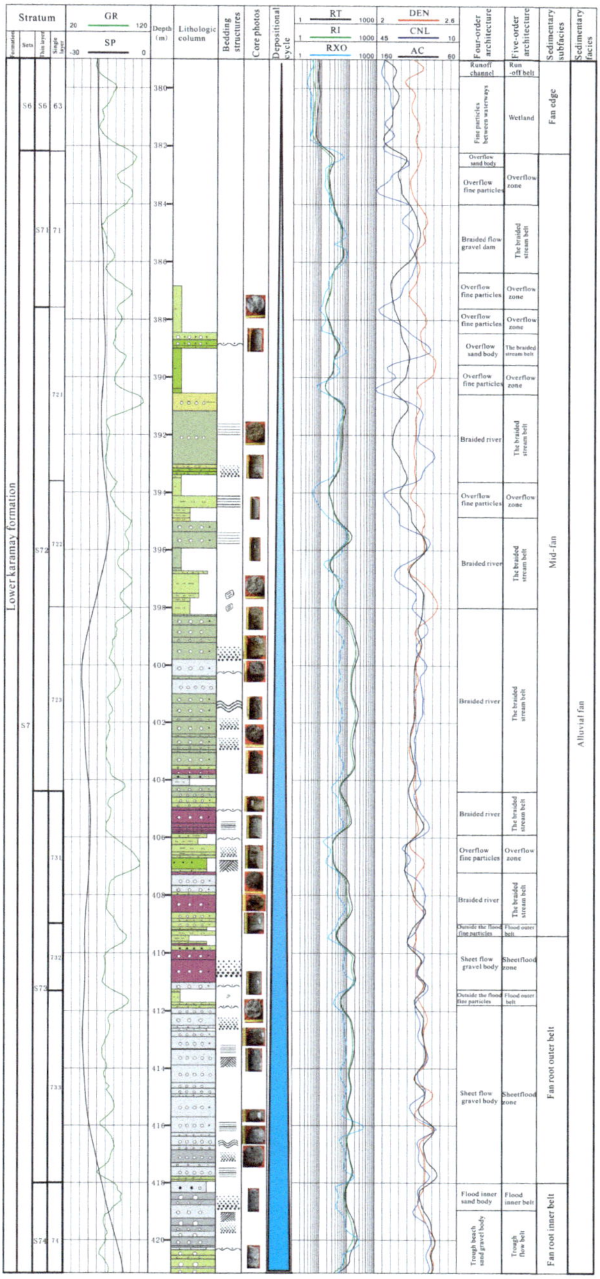

Fig. 5.10 Single well comprehensive histogram of a well in Kexia Formation of a certain area in the northwestern margin of Junggar Basin

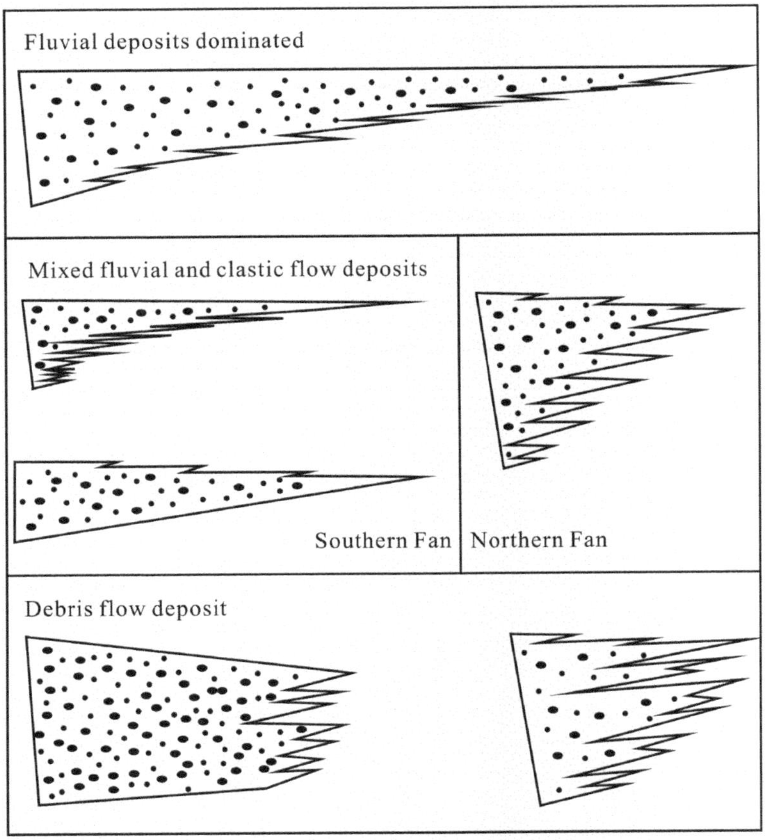

Fig. 5.11 The influence of different sedimentation on fan morphology Mei Zhichao (1994)

5.3 Lithofacies Characteristics of Volcanic Gas Reservoirs and Their Effects on Physical Properties

Volcanic oil and gas reservoirs have witnessed a history of 120 years of exploration abroad. With the increasingly challenging situation of oil and gas exploration and development in clastic and carbonate rocks globally, and the sharp increase in oil and gas consumption owing to the rapid growth of the social economy, the exploration and development of volcanic oil and gas reservoirs has gradually aroused researchers' interests. The broad resource prospects make volcanic oil and gas reservoirs a research hotspot. The type of volcanic facies is closely related to the characteristics of reservoir development, and the volcanic facies analysis is crucial for predicting favorable reservoirs (Tang et al. 2004; Kawamoto et al. 2000; Xu et al. 2006, 2008; Shu et al. 2007; Huang et al. 2006; Wang et al. 2008; Sun et al. 2009; Zhao et al. 2008; Feng et al. 2014; Liu et al. 2018). In this section, the lithofacies of volcanic gas reservoirs in the Xudong area of the Songliao Basin's Xushen gas field are studied to provide a geological basis for the effective development of the gas field. The information on the study area and the target horizons are detailed in Sect. 5.2, Chapter 2.

5.3.1 *Volcanic Rock Types*

Volcanic lithology varies with volcanic facies. Therefore, accurately identifying and dividing volcanic lithologies has become critical in analyzing volcanic facies. Wang et al. (2007a, b, c) reviewed the geophysical technique, cross-plotting technique in logging, principal component analysis, elemental capture spectroscopy logging technique, image-logging technique, BP neural network in artificial intelligence techniques, and self-organizing map (SOM) neural network technique, and proposed that logging technique is the best way to obtain reservoirs' physical property parameters. Luo et al. (2008) identified the lithofacies of the Yingcheng Formation volcanics in the Shengping Gas Field of the Songliao Basin based on the lithology division of volcanic rocks and the characteristics of volcanic eruptions. The volcanic reservoirs of the Ying-1 Member in the Xudong area of the Xushen Gas Field were formed from multiple volcanic eruptions, comprising various volcanic rocks. To identify the lithology of volcanic rocks, core and microscopic thin section observations were primarily conducted for some critical wells (Fig. 5.12), and logging lithology interpretation was conducted. The results show that the volcanic rocks in the Ying-1 Member can be divided into 10 lithologies (rhyolite, basalt, rhyolitic breccia lava, rhyolite tuff lava, rhyolitic welded breccia, rhyolitic welded tuff, rhyolitic tuff, rhyolitic volcanic breccia, sedimentary tuff, and sedimentary volcanic breccia) in four categories such as volcanic lava and volcaniclastic rocks (Fig. 5.12). The rhyolite, rhyolitic tuff, and sedimentary volcanic breccia are the most developed.

5.3.2 *Identification of Volcanic Massif*

Volcanic massif identification has always been an important subject in volcanic reservoir studies (Kawamoto et al. 2000; Tang et al. 2007). Tomohisa et al. (2000) identified and tracked volcanic massifs for the stratigraphic division of volcanic reservoirs when building the geological model of heterogeneous volcanic reservoirs by using the method of petrology. Under the joint control of the crater location and volcanic eruption cycle, many volcanic massifs of different shapes and sizes are formed in the same volcanic eruption period. These volcanic massifs are intersected and superimposed, with various volcanic facies developed at different parts. The volcanic facies jointly constitute volcanic reservoirs. Since the crater controls the location and scale of volcanic massif (also coupling with the structural factors such as topography), in this book, verification and comparison with well and seismic data are

Fig. 5.12 Lithological core and microscopic characteristics of volcanic rocks in the first member of Yingcheng Formation in Xudong area, Songliao. Basin A-stomatal rhyolite, XS14 well, depth 3784.85–3785.53 m; B-breccia lava, XS21-1 well, depth of 3752.21–3752.68 m; C-sedimentary volcanic breccia, XS21, depth 3836.62–3836.76 m; D-fused tuff, Xushen 231, depth 3760.00–3760.13 m; E-crystal tuff, XS42, depth 3702.06; F- Volcanic breccia, XS24, depth 3683.52 m

conducted to find the crater location. Then, considering that volcanic massifs of the same period exhibit similar seismic reflections and the superimposed boundaries of volcanic massifs show strong seismic reflections or distinct changes in different seismic wave groups (Fig. 5.13), the volcanic massifs are identified according to the lithology and electricity of the volcanic rocks. On this basis, volcanic facies are identified depending on the changes in the lower-level seismic reflection characteristics. With the support of Petrel software, and using the well and seismic data, the volcanic massifs in the Ying-1 Member in the Xushen Gas Field are identified and tracked in detail (Fig. 5.14), laying a solid geological foundation for volcanic facies analysis.

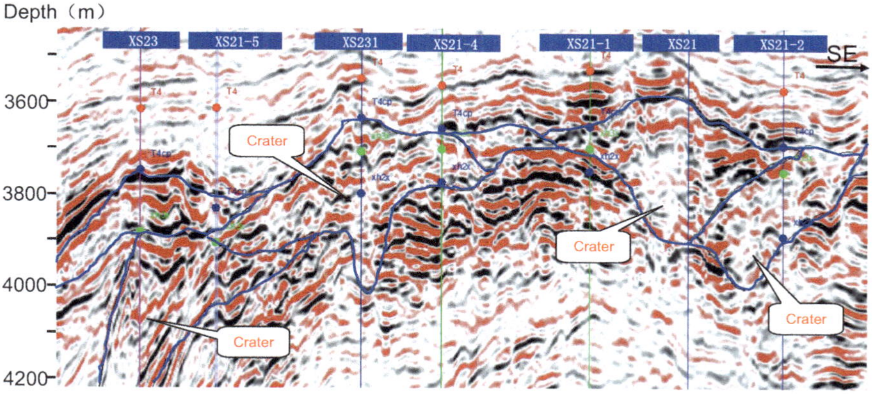

Fig. 5.13 Characteristics of seismic reflection of volcanic reservoir of one Member of Yingcheng Formation in Xushen Gas Field

5.3.3 Classification and Special Distribution of Volcanic Facies

5.3.3.1 Classification of Volcanic Facies

Volcanic facies refer to the synthesis of various rocks formed owing to volcanism under specific geological conditions. Each volcanic facies has its own typical volcanic rock types and associations. Therefore, identification and division of volcanic lithologies becomes an effective method for volcanic facies analysis. Many scholars have devoted great efforts to volcanic facies identification and division (Wang et al. 2008; Sun et al. 2009; Luo et al. 2008; Guo et al. 2006; Yu et al. 2009). In addition, the geophysical methods play an important role in the division of volcanic facies. Guo et al. (2006) analyzed the relationship between lithofacies and electrofacies of the volcanic rocks in northern Songliao Basin and accordingly used electrofacies to identify the lithofacies of volcanic rocks. Li et al. (2009) effectively predicted the spatial distribution of volcanic rocks and the favorable facies belt of volcanic reservoirs in the Ying-1 Member in northern Songliao Basin by using seismic facies and electrofacies. Yu et al. (2009) predicted the distribution of volcanic facies in the Yingcheng Formation in the Anda fault depression of the Songliao Basin, according to the seismic reflection characteristics of volcanic facies, together with the coherence technique and amplitude attributes, and discussed the application of the seismic technique to identifying volcanic facies and physical properties of volcanic reservoirs, showing good results.

Here, based on the rock types and the log, core and test data, the lithofacies are divided and identified for single wells. Then, typical sections in the study area

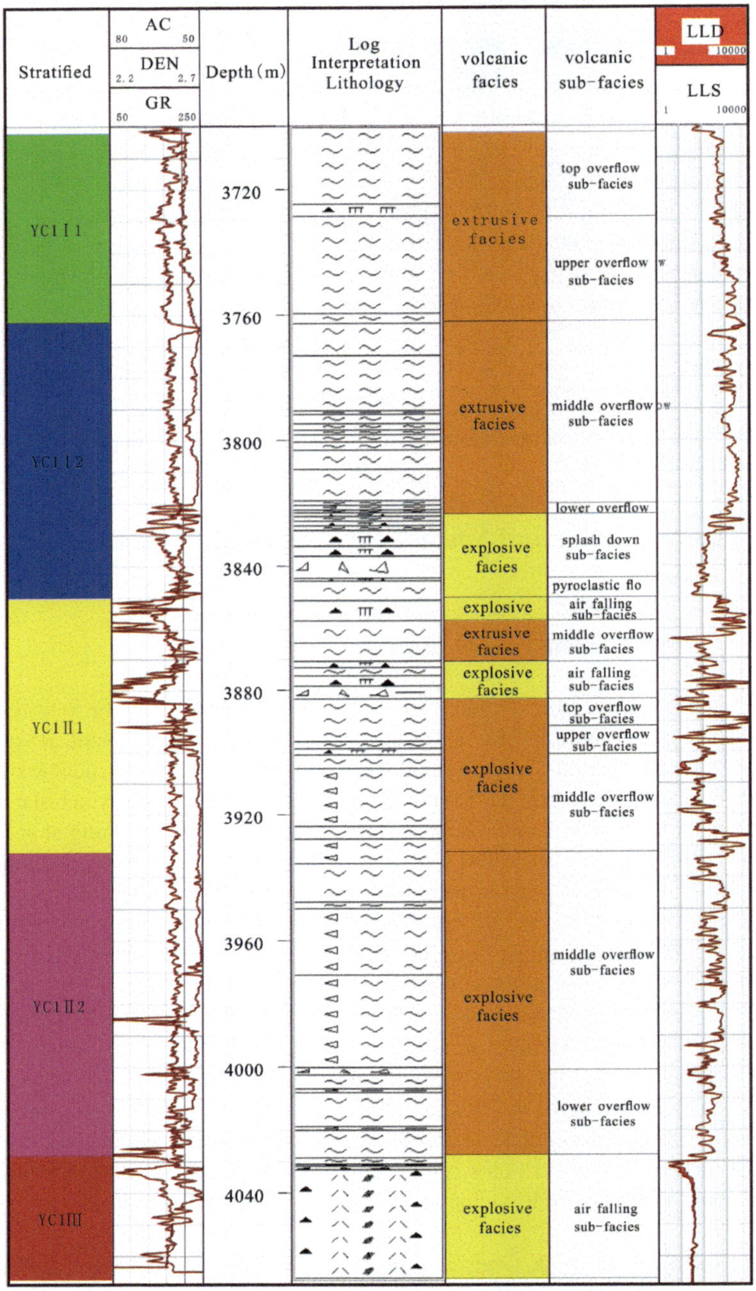

Fig. 5.14 Characteristics of electric logging curve of volcanic facies in well XS-A of the first member of Yingcheng Formation in Xudong area, Songliao Basin

are selected for well-seismic correlation, in order to analyze the spatial distribution of volcanic facies. Next, using the Petrel software and a computer workstation, seismic attributes that can reflect the information of lithofacies are extracted for analyzing the seismic facies of volcanic rocks. Finally, the seismic facies are converted into volcanic facies to describe the plane distribution of different types of volcanic facies. Based on the volcanic facies model, combined with core, logging, and seismic data, the volcanic rocks are divided into 16 subfacies in five facies, i.e. volcanic conduit (4.55%), extrusive (1.35%), explosive (44.71%), effusive (43.40%), and volcanogenic sedimentary (5.99%), as shown in Table 5.1. The data in parentheses represent the thicknesses percentage of volcanic facies obtained from the fine logging interpretation of 32 wells.

(1) Volcanic conduit facies

The volcanic conduit facies is in the lower part of the entire volcanic edifice, and formed during the entire volcanic cycle and later stages. It can be divided into the volcanic neck, subvolcano, and cryptoexplosive breccia subfacies (Table 5.1). The volcanic conduit facies is just below the top of the volcanic cone, with a near-upright, columnar occurrence. It shows intermittent and chaotic internal seismic reflections (Fig. 5.13). It shows serrated shape with high amplitude on the electric log(Fig. 5.14).

(2) Extrusive facies

The extrusive facies in the crater's upper part represents a dome morphology, and is formed in the late stage of volcanic cycle. It can be divided into intrazone, middle zone, and exozone subfacies (Table 5.1). The intrusive facies shows discontinuous reflections on the seismic section, with the events radiating from the bottom (Fig. 5.13). It is usually developed near the volcanic conduit facies. The intrusive facies shows a microserrated shape with high amplitude on the electric log(Fig. 5.14).

(3) Explosive facies

The explosive facies is formed in the early stage of volcanism, and can be divided into the fallout, thermal pyroclastic flow, thermal base surge, and fallout subfacies (Table 5.1). The explosive facies often shows a hummocky reflection on the seismic section, mostly chaotic in the interior, with strong reflection at the top and weak reflection in the interior (Fig. 5.13). It shows medium and low values, and serrated shape on the electric log(Fig. 5.14).

(4) Effusive facies

The effusive facies is formed in the early–middle stage of volcanic eruption cycle. It is generated when the magma containing crystals and contemporaneous breccia flows along the surface under the combined action of the subsequent ejecta and its own gravity (Table 5.1). The Ying-1 Member in the Xudong area of the Xushen gas field is dominated by effusive facies, which can be found in acidic, neutral, and basic volcanic rocks and can be divided into lower, middle, upper, and top subfacies. The effusive facies shows intermittently continuous moderate–strong reflections on the seismic section (Fig. 5.13), and thick-layered and microserrated shape with medium–high values on the electric log(Fig. 5.14).

Table 5.1 Volcanic facies in the Ying-1 Member, Xudong area, Songliao Basin (modified after Wang et al. 2008)

Facies	Subfacies	Genetic mechanism and division mark	Rock type	Texture	Digenesis
Volcanic conduit (in the lower part of the volcanic massif)	Volcanic neck	The lava flow stagnates and fills the volcanic conduit, and crater collapses form the fillings	A transitional type between volcaniclastic rock and lava; mostly welded breccia and breccia lava, as well as breccia crystal tuff	Porphyritic texture, resorption texture, breccia tuff texture	Magmatic melt condensing solidification, and magma solidification with various breccia and tuffs
	Subvolcano	It is formed by the condensation and crystallization of magma intruding into the surrounding rock			Magma condensation and crystallization
	Cryptoexplosive breccia	When the volatile-rich magma intrudes into the rock fracture zone, some pressure is released but not completely, causing underground explosions			Cementation with rock juices (hydrothermal minerals) or fine clasts which have the same or different composition as breccia
Extrusive (mostly formed in the early and late stages of volcanic eruption cycles)	Intrazone	The high-viscosity magma flows under the compression of internal force, piles near the crater to form a dome or condenses at the lava front, and then deforms, scrapes and wrap the young and old rock fragments	Autobrecciated lava, lava	Welded breccia and sintered tuff textures, vitreous and perlitic textures, porphyritic texture, mortar texture	Solidification and condensation of melting magma
	Mesozone				
	Exozone				Magma condensation, and welding of young and old rock fragments and clasts

(continued)

Table 5.1 (continued)

Facies	Subfacies	Genetic mechanism and division mark	Rock type	Texture	Digenesis
Explosive (mostly formed in the early stage of volcanic eruption cycle)	Avalanche	When the magma rises near the crater, the surrounding rock materials carried by it and the material of the lava fall and accumulate nearby	Breccia lava, tuff lava, welded breccia, etc	Welded breccia tuff texture	Compaction
	Thermal pyroclastic flow	After the emanation column collapses, the hot debris flows along the surface under the combined action of subsequent ejecta and its own gravity. The rock type is mostly welded tuff containing crystal, glass, magma, and debris	Rhyolite crystal welded tuff	Welded tuff texture, pyroclastic texture	Magma condensation and cementation + compaction
	Thermal base surge	Under the action of gravity, the gas–solid–liquid multiphase turbidity flow system in emanation is transported by suspension near the surface, and the debris-carrying vapor flows. It is mostly tuff containing crystal, glass, and magma	Rhyolitic crystal tuff	Pyroclastic texture (mainly crystal tuff texture)	Compaction

(continued)

Table 5.1 (continued)

Facies	Subfacies	Genetic mechanism and division mark	Rock type	Texture	Digenesis
	Fallout	Solid and plastic ejecta in emanation fall freely (under the action of wind). It is represented by agglomerate, breccia and crystal tuff containing volcanic bombs and pumice fragments	Volcanic breccia, crystal lithic tuff, rhyolitic breccia crystal tuff, and rhyolitic crystal tuff	Agglomerate texture, breccia texture, tuff texture	Compaction
Effusive (mostly formed in the middle stage of volcanic eruption cycle)	Top	The magma containing crystals and contemporaneous breccia flows along the surface under the combined action of the subsequent ejecta and its own gravity. Primary voids are developed in the upper part, structural fractures are developed in the lower part, and both are underdeveloped in the middle part	Stomatal rhyolite, basalt, and dacite	Spherulitic texture, fine-crystal texture	Magmatic melt condensing solidification
	Upper				
	Middle		Basalt, rhyolite, and dacite	Fine-crystal texture, porphyritic texture	
	Lower		Fine-crystal rhyolite, basalt, and dacite	Vitreous texture, fine-crystal texture, porphyritic texture, breccia texture	
Volcanogenic sedimentary (formed in any stage of volcanic eruption cycle)	Retransported	Pyroclastics are transformed by water flow	Various volcanic breccia and tuffs	Terrigenous clastic texture	Compaction
	Epiclast-bearing	It is mainly pyroclasts, possibly with other terrigenous clastic materials	Mixture of volcanic rocks and sedimentary rocks	Terrigenous clastic texture	Compaction

(5) Volcanogenic sedimentary facies

The volcanogenic sedimentary facies usually coexists with volcanic rocks and may appear in all stages of volcanic activity. Its fragments contain a large content of volcanic debris, and present mainly as clastic deposits containing volcanic materials between volcanic domes. The volcanogenic sedimentary facies can be divided into epiclast-bearing and retransported subfacies (Table 5.1). It shows moderate–strong, continuous, and stable reflections on the seismic section (Fig. 5.13), and rhythmic characteristics and varying thicknesses on the electric log(Fig. 5.14).

5.3.3.2 Spatial Distribution of Volcanic Facies

The lithofacies division of a single well shows that the explosive facies and effusive facies are apparently dominant in thickness, as 44.71% and 43.40%, respectively, while the volcanogenic sedimentary, volcanic conduit, and extrusive facies take less proportion of thickness, as 5.99%, 4.55%, and 1.35%, respectively (Fig. 5.14). It can be seen that different types of volcanic subfacies are developed in the same volcanic facies, but they do not exhibit obviously complete subfacies vertically, owing to the interactive destruction between volcanic massifs formed in different periods or by different volcanic eruption cycles in the same period.

As observed from the section of volcanic facies (Fig. 5.15), the extrusive and explosive facies are generally adjacent to the volcanic conduit facies, and the effusive facies is spatially farther from the volcanic conduit facies. The later volcanic massif is strongly destructive to the early volcanic massif, and the early volcanic massif may change abruptly at the interface with the later volcanic massif. Due to the volcanic activities in different periods, the volcanic conduit and extrusive facies extend laterally by one well spacing (2.5–4.0 km), while the explosive, effusive and volcanogenic sedimentary facies extend farther—possibly by two well spacings or longer locally. Different volcanic facies are spatially superimposed, forming different volcanic edifices. The distribution of volcanic facies is jointly controlled by the factors such as the intensity of volcanic activity and topography. Among the subfacies types, the fallout subfacies of the explosive facies and the middle subfacies of the effusive facies are best developed. Different volcanic eruption cycles are developed within different volcanic massifs (volcanic edifices). For the volcanic conduit facies, the volcanic neck subfacies is dominant; for the extrusive facies, the mesozone subfacies is dominant; for the explosive facies, the fallout and avalanche subfacies are dominant; for the effusive facies, the middle subfacies is dominant; for the volcanogenic sedimentary facies, which is limited in distribution, the epiclast-bearing and retransported subfacies are equally matched. Due to the factors such as the intensity of volcanic eruption and lithology, the volcanic subfacies within volcanic massifs vary rapidly in thickness vertically, but extend in a limited range laterally. Note that the section does not cross the crater's center, the hummocky appearance of the crater (at Wells XS21-5, XS21, XS21-2, etc.) is not evident on the section, but a part of the flanks is exhibited.

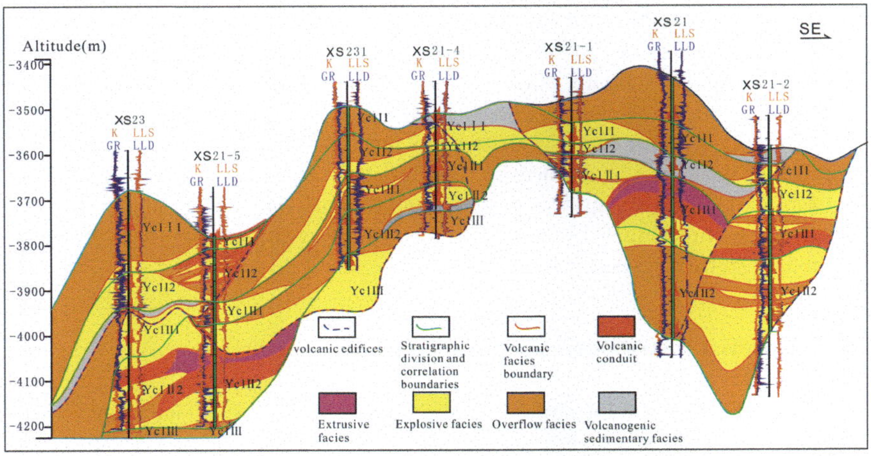

Fig. 5.15 Volcanic facies profile of the first member of Yingcheng Formation in Xudong area, Songliao Basin

The volcanic eruption cycle III of the Ying-1 Member in the XS12 well block is taken as an example to describe the plane distribution of the volcanic facies (Fig. 5.16). The XS12 well block is located in the southwest of the study area, where three major craters can be identified near Wells XS14, XS141, and XS12. The explosive and effusive facies are dominant, which is consistent with the results of single-well volcanic facies division. They are distributed in near N-S direction; the explosive facies is concentrated in the southwestern and northeastern parts of the well block, and the effusive facies is concentrated in the central part of the well block, covering a larger area than the explosive facies. The comparison of the fault location with the crater location shows that the crater and volcanic facies are significantly controlled by faults, and the fissure-central eruption pattern is dominant in the well block.

5.3.3.3 Influence of Volcanic Facies on Physical Properties of Reservoirs

The physical properties of volcanic rocks are closely related to their lithologies and lithofacies. Lithology is the material basis at which other factors affect the physical properties of volcanic reservoirs. Lithologies vary significantly in the attributes such as hardness, density, composition, texture and structure, resulting in volcanic rock types with different physical properties, such as porosity and permeability (Wang et al. 2008). Various lithologies are developed in different volcanic facies formed by volcanic eruption cycles. Therefore, volcanic facies and subfacies control the physical properties and storage space types/changes of volcanic reservoirs. The reservoirs of different facies or of the same facies but different subfacies may differ considerably in characteristics. Therefore, the physical properties of volcanic reservoirs can be

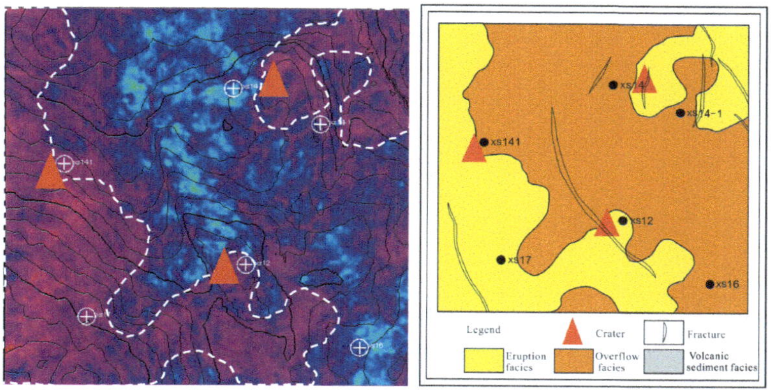

Fig. 5.16 Seismic facies and sedimentary facies plan of volcanic cycle III in the first member of Yingcheng Formation, XS12 area, Xudong area, Songliao. Basin The seismic facies plane diagram of volcanic cycle III in the A-XS12 well area; the sedimentary facies plan of volcanic rock cycle III in B-XS 12 well area

analyzed based on research on volcanic facies to identify the relationship between volcanic facies/subfacies and physical properties of reservoirs, thereby guiding the exploration and development of volcanic gas reservoirs.

The single-well volcanic facies are correlated with the data of 206 cores taken from the study area (Figs. 5.17 and 5.18). Two parameters (effective porosity and total permeability) are adopted to illustrate the physical properties of volcanic reservoirs of different facies. It can be seen that the explosive facies corresponds to the best physical properties, with an average effective porosity of 7.75% and an average total permeability of 3.361 mD, followed by the extrusive facies, volcanic conduit facies and effusive facies, as well as the volcanogenic sedimentary facies. In Fig. 5.18, the avalanche subfacies of the explosive facies corresponds to the best physical properties, with an average effective porosity of 9.00% and an average total permeability of 4.793 mD, followed by the exozone and mesozone subfacies of the extrusive facies, the lower, top and upper subfacies of the effusive facies, the volcanic neck subfacies of the volcanic conduit facies, and the fallout subfacies of the explosive facies; the thermal pyroclastic flow subfacies of the explosive facies, the subvolcano subfacies of the volcanic conduit facies, the middle subfacies of the effusive facies, the epiclast-bearing subfacies of the volcanogenic sedimentary facies, the thermal base surge subfacies of the explosive facies, and the intrazone subfacies of the extrusive facies reflect the worst physical properties. By analyzing the volcanic facies, the location of favorable reservoirs can be predicted to provide a reference for the effective development of volcanic gas reservoirs (Chen et al. 2011a, b).

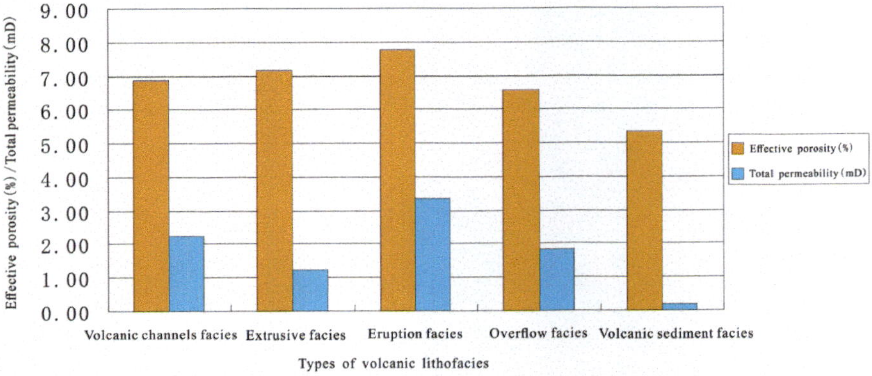

Fig. 5.17 Columnar diagram of the relationship between volcanic facies and reservoir physical properties of the first member of Yingcheng Formation in Xudong area, Songliao Basin

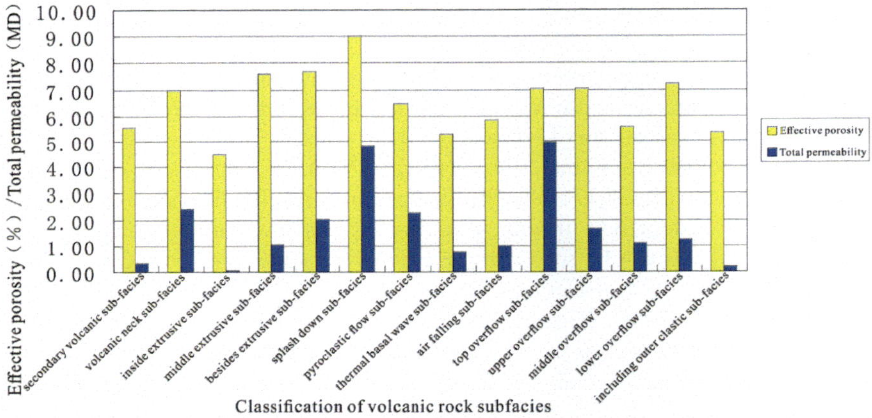

Fig. 5.18 Columnar diagram of the relationship between volcanic subfacies and reservoir physical properties of the first member of Yingcheng Formation in Xudong area, Songliao Basin

5.4 Geological Genetic Characteristics of Permeability Barriers in Thermally Recovered Heavy Oil Reservoirs

A permeability barrier is a rock mass blocking fluid flow in the reservoir system (Wu 2010). Wu et al. (1999) studied the flow units in the braided river delta reservoirs of the Middle Jurassic Sanjianfang Formation in the Wenjisang–Mideng Oilfield, the Turpan–Hami Basin, and indicated that the flow units can be studied at two levels—the delineation of the distribution of connected sandbodies and permeability barriers as the first level, which has a significant effect on later studies of flow units and permeability barriers. Baosheng et al. (2004) divided the pore structure and flow units in the reservoirs in the Dong III Formation in the Tangjiahe Developing Area of the Beidagang Oilfield, the Huanghua Depression, by analyzing the microscopic pore structure, permeability barriers, and connected bodies. Dipple et al. (2005) investigated the effect of quartz veins as permeability barriers on regional fluid flow, with a case study of the Mesoproterozoic in the Mt Painter Block, South Australia. Li et al. (2007) corresponded the reservoir fluid flow barriers to the fourth-order architecture boundary. Peng et al. (2009) analyzed the depositional, diagenetic, tectonic, and production performance factors affecting the permeability barrier distribution in the complex fault-block reservoirs in the Shusan District of Liaohe Oilfield. Shu et al. (2012) divided the flow units of the braided river delta front reservoirs in the $E_3{}^1$ oil reservoir of the Gasikule Oilfield based on the research of permeability barriers. Mohammadi et al. (2013) analyzed the impact of discontinuous impermeable barriers on sweep efficiency in a pore-level screening study on miscible/ immiscible displacements in heterogeneous models. Previous studies on permeability barriers are mostly conducted using simple correlation analysis in reservoir flow unit division. However, there have been no special works on permeability barriers, especially those in thermally recovered heavy oil reservoirs. The permeability barriers have great impacts on the steam advancement during the steam flooding thermal recovery process of heavy oil reservoirs, so the research on the types and spatial distribution of permeability barriers is of practical significance to the steam-flooding thermal recovery of heavy oil reservoirs. The information on the study area and the target horizons are detailed in Sect. 5.2, Chap. 3.

5.4.1 Classification and Identification of Permeability Barriers

The classification of permeability barriers has always been the foundation and essential content of researches on permeability barriers. Many scholars have devoted efforts in this aspect. Wu et al. (1999) divided permeability barriers into mudstone, cemented zone, and sealing fault, and introduced their respective development characteristics. Dou et al. (2004) divided the permeability barriers in the Paleogene fluvial reservoirs in the Gangdong area of the Northern Dagang Oilfield into mudstone, calcareous cement, and sealing fault. Zhang (2006) divided the permeability barriers in the Ordovician reservoirs in the Tahe Oilfield into permeability barriers filled with tight carbonate rock matrix, permeability barriers formed by filling, and permeability barriers formed by sealing faults. Shen et al. (2007) divided the permeability barriers in the Upper Karamay Formation in the Liuzhong area of the Karamay Oilfield into fault, physical property, and unconformity permeability barriers. Chen et al. (2007) considered that the geobodies separated by permeability barriers are sandbodies, and divided the permeability barriers into barrier, interlayer, and fault. Depending on rock types, the reservoirs have their physical properties sensitive to different factors, such as tectonism, sedimentation, and diagenesis; therefore, the classification of permeability barriers is different for such reservoirs as clastic, carbonate or igneous rocks. Currently, working from the perspective of genesis predominates the classification of permeability barriers.

5.4.1.1 Classification of Permeability Barriers

As mentioned above, a permeability barrier is a rock mass blocking fluid flow in the reservoir system. Here, according to the availability of data and the characteristics of permeability barrier development, and using the core, SEM, thin section, logging and test data, the permeability barriers in the Yulou reservoir in the Liaohe Basin's West Sag are divided by genesis into: sedimentary permeability barriers (composed of fine-grained sediments, such as underwater distributary interchannel mud of sedimentary origin), diagenetic permeability barriers (formed by diagenetic processes, such as compaction and cementation), and sealing fault permeability barriers, with the sedimentary permeability barriers in dominance.

(1) Sedimentary permeability barriers

The sedimentary permeability barriers, controlled by sedimentation, are primarily composed of underwater distributary interchannel sand or underwater distributary interchannel mud, and lithologically include siltstone, argillaceous siltstone, silty mudstone, and mudstone (Fig. 5.19), with the silty mudstone and mudstone in dominance. The barriers are formed by the deposition of fine-grained suspended materials due to the weakening hydrodynamics. They mainly exist in two genetic types. One is the argillaceous layer between sandy laminae formed by the changes in hydrodynamic conditions resulting from diversions of underwater distributary channels, which is generally thin and dominantly occurs in the form of an interlayer. The other is the argillaceous layer formed in the drought period during lake flooding or in the quiet-water period, which is generally thick and often appears as the barrier. Restricted by the distribution scale, the interlayer has significantly less influence on the fluid flow in the reservoir than the barrier. The sedimentary permeability barriers can be identified using various data (e.g. core and logging), and detailed analysis of sedimentary facies/microfacies is an effective method. The sedimentary permeability barriers demonstrate low and flat electric logging curves, with small or slight relief (Fig. 5.20). The sedimentary permeability barriers are the most developed in the Yulou reservoir, and play a decisive role in controlling the reservoir connectivity.

(2) Diagenetic permeability barriers

Diagenesis exerts a critical influence on the physical properties of reservoirs (Wu et al. 1998; Ji 2009). The diagenetic permeability barriers in the Yulou reservoir are controlled by diagenesis, including cementation and compaction, which negatively affect the reservoirs, that is, deteriorate the physical properties of the reservoirs by reducing the porosity, thus forming permeability barriers.

Fig. 5.19 Core characteristics of sedimentary seepage barrier of Yulou oil layer in a certain area of western sag of Liaohe Basin. A-JJ2,1003.74–1003.79 m, silty mudstone, sedimentary seepage barrier, underwater interdistributary mud; B-J23-261,975.3–975.35 m, gray-brown siltstone, oil spots, wavy bedding, sedimentary seepage barrier

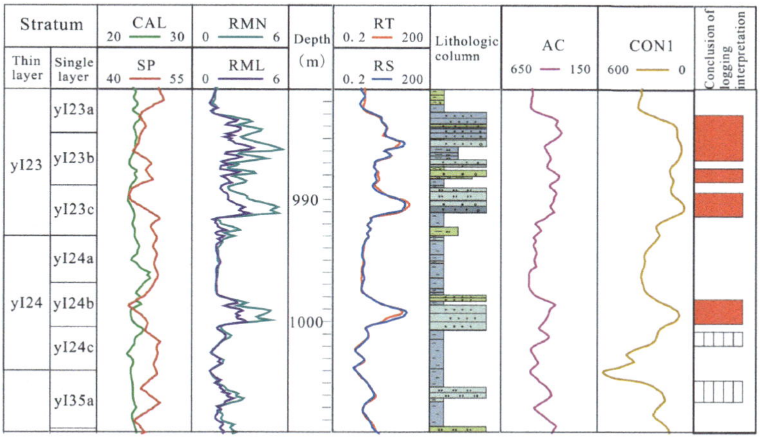

Fig. 5.20 Sedimentary seepage barrier characteristics of Yulou oil layer (depth: 992–998 m) of well J91 in a certain area of western sag in Liaohe basin)

Overall, the reservoir is weakly compacted—mainly by way of cementation of carbonate, clay minerals, and silica. The carbonate cement is mainly tight clastic rock formed by the metasomatism of carbonates that are generated from the combination of abundant CO_2 released by the thermal evolution of organic matters and Ca^{2+}, Mg^{2+} and other cations in the formation water, along with the rising temperature and pressure owing to the increasing burial depth during the diagenesis. The siliceous cement is mainly authigenic quartz formed in the intergranular pores by chemical precipitation. In the Yulou reservoir, the diagenetic permeability barriers were formed by the interactive transformation of various clay minerals during reservoir development to block the pores. The siliceous cement manifests as feldspar overgrowth (Fig. 5.21). The calcareous cement can be identified through the distinct calcareous tip on the resistivity log

(Fig. 5.22). Generally, the diagenetic permeability barriers are underdeveloped in the Yulou reservoir, especially the permeability barriers formed by calcareous cement, which can only be seen in some yII sublayers (Fig. 5.22). Moreover, they are thinner than 1 m, and extend in a short distance spatially. Therefore, the diagenetic permeability barriers have limited influences on production practices.

(3) Sealing fault permeability barriers

As a critical technique, fault sealing is analyzed by using the production performance data or fluid properties on the fault walls (Zhang et al. 2011; Li et al. 2007). Here, the sealing fault permeability barriers are investigated by comparing the water analysis data of the fault walls. Four primary faults (F1, F2, F3, and F4) occur in the study area. The water analysis data of 10 wells in five pairs are selected for comparison to judge the fault orders (Table 2.2). It is found that the faults F1 and F2 belong to the third-order sag-controlling faults, and F3 and F4 belong to the fourth-order faults. The analysis data of formation water in the hanging and foot walls of faults are compared to evaluate the fault sealing

Fig. 5.21 Diagenetic seepage barrier characteristics of Yulou oil layer in a certain area of western sag of Liaohe Basin. A-J2, kaolinite clay to illite conversion, diagenetic seepage barrier, 997.02 m, × 2400; B-J2, feldspar overgrowth grade I, diagenetic seepage barrier, 990.18 m, × 6000; C-J2, particle point-line contact, compaction, 954.04 m, diagenetic seepage barrier, × 100; D-J93,1033.84–1033.94 m, medium sandstone, siliceous cementation, diagenetic seepage barrier

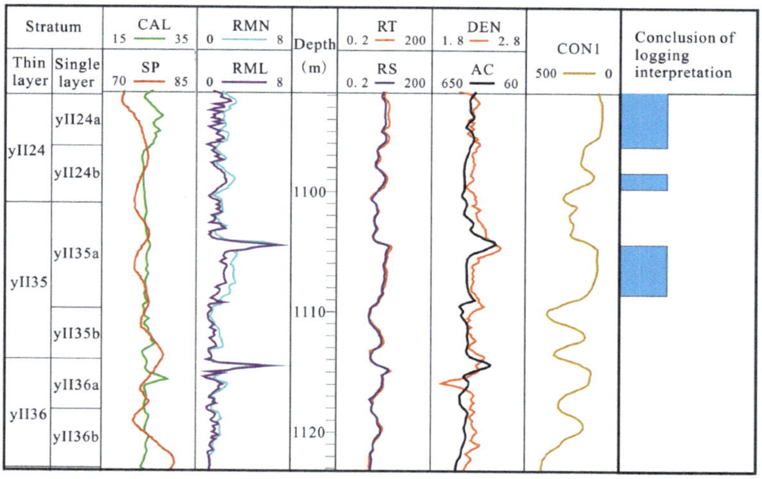

Fig. 5.22 Characteristics of diagenetic seepage barrier in well J10-22 of Yulou oil layer in a certain area of western depression of Liaohe Basin (depth: 1104-1105 m; 1114-1115 m)

and then delineate the sealing fault permeability barrier. If the comparison shows great difference, it is suggested that the fault sealing is good to form the sealing fault permeability barrier. If the comparison shows similarity, it is suggested that the fault sealing is poor to form the sealing fault permeability barrier. For fault F1, the water analysis results of Wells A1 and A2 on the hanging and foot walls are compared, showing huge differences, particularly in the contents of Mg^{2+}, Ca^{2+}, $SO_4{}^{2-}$ and CO_3, which indicates that the fault is sealing. The comparison of fractures C1 & C2 and E1 & E2 as two groups reveals that the indicators, such as $K^+ + Na^+$, Ca^{2+}, CO_3, HCO_3, and total hardness, are identical or similar; hence, it can be concluded that the fluids between the hanging and foot walls are connected, and the fault is not sealing and thus cannot form a sealing fault permeability barrier. The comparison of fractures of B1 & B2 and D1 & D2 as two groups demonstrates that the indicators, such as Mg^{2+}, $SO_4{}^{2-}$, CO_3, total salinity, and total alkalinity, are very different; hence, it can be concluded that the fluids between the hanging and foot walls are disconnected, and the fault is sealing and can form a sealing fault permeability barrier. Among the faults in the study area, F1 and F4 are sealing fault permeability barriers.

5.4.1.2 Spatial Distribution of Permeability Barriers

(1) Sectional distribution

Affected by the frequent conversion of underwater distributary channels at the fan delta front, the distribution of sedimentary permeability barriers in the Yulou reservoir varies significantly. Usually, the sedimentary permeability barrier is relatively developed at the underwater distributary interchannel mud and underwater distributary interchannel sand, but not developed at the underwater distributary channel and river-mouth bar (Fig. 5.23). Under the joint control of the provenance and the fan delta front's sedimentary cycle, the permeability barriers in the upper and lower parts of the Yulou reservoir are characterized by great thickness, large extension distance, and small numbers; the permeability barriers in the middle part are small in scale but appear frequently in large numbers. Moreover, in the NW–SE direction, as the distance to the provenance increases, the sediment grains gradually become smaller, and the permeability barrier development is gradually intensified. These characteristics should be fully considered when implementing separate-layer steam injection.

(2) Plane distribution

Here, the plane distribution of sedimentary permeability barriers, which dominate the Yulou reservoir, is described from the sedimentary perspective. The sedimentary permeability barriers can be subdivided into barriers and interlayers (Figs. 5.24 and 5.25), of which the former plays a more significant role than the latter in field development. Taking the barrier between yI_3^{5b} and yI_3^{5c} as an example, the sedimentary permeability barriers vary greatly in thickness on the plane from 0 to 5 m. Affected by the sediment source, the sedimentary permeability barriers with different thicknesses are roughly distributed in form of NW–SE strips. The permeability barriers see their thick parts primarily at the underwater distributary interchannel mud, and 0 thickness at the position where underwater distributary channels are overlapped. The barriers between other sublayers show similar development characteristics, but their specific positions are different owing to the diversion of underwater distributary channels.

It shows the frequency distribution of interlayers in yI_1^{1b} (Fig. 5.25). It can be seen that the interlayers with different frequencies are generally distributed in form of NW–SE strips on the plane, which is consistent with the distribution pattern of barriers. However, the interlayers are significantly smaller than the barriers, and extend for a limited distance spatially. The other sublayers basically exhibit the similar distribution of interlayers to yI_1^{1b}.

When shifting from steam huff-and-puff to steam flooding, the injector-producer pattern in the steam flooding should be adjusted and deployed with a full consideration to the plane distribution of permeability barriers, so as to ensure the connectivity between injectors and producers for a good flooding effect.

5.4.1.3 Influence of Permeability Barriers on Development of Thermally Recovered Heavy Oil Reservoirs

The permeability barriers significantly control fluid flow in the reservoirs. However, not all permeability barriers can effectively hinder the fluid flow. In other words, only the permeability barrier with a sufficient thickness and extension can block the steam injected, while the thin permeability barrier just slightly changes the steam advance

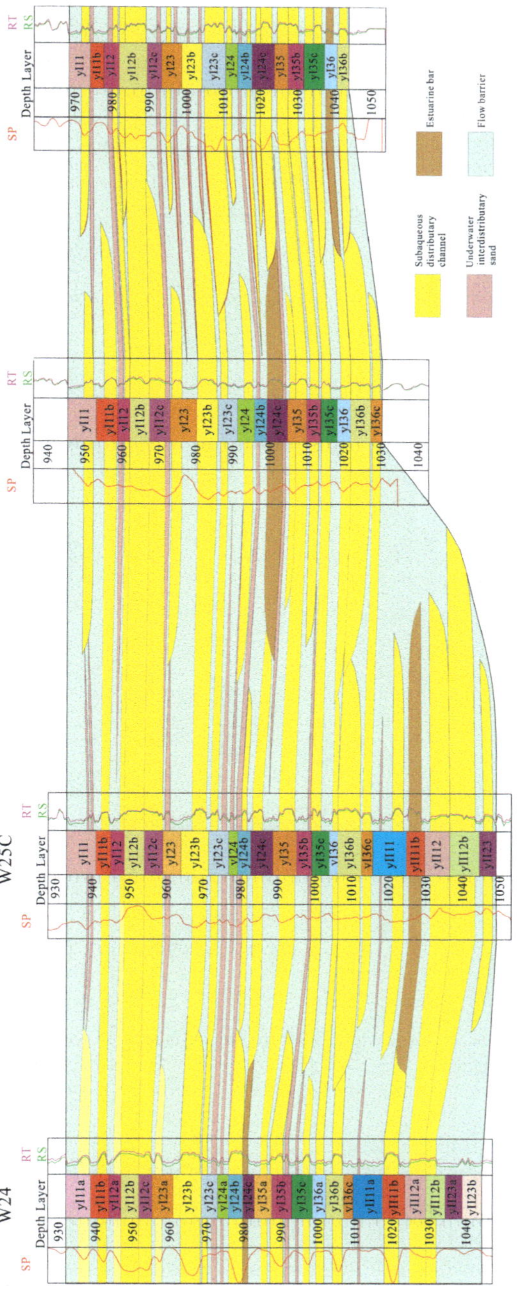

Fig. 5.23 Development characteristics of seepage barrier profile of Yulou oil layer in a certain area of western sag of Liaohe Basin

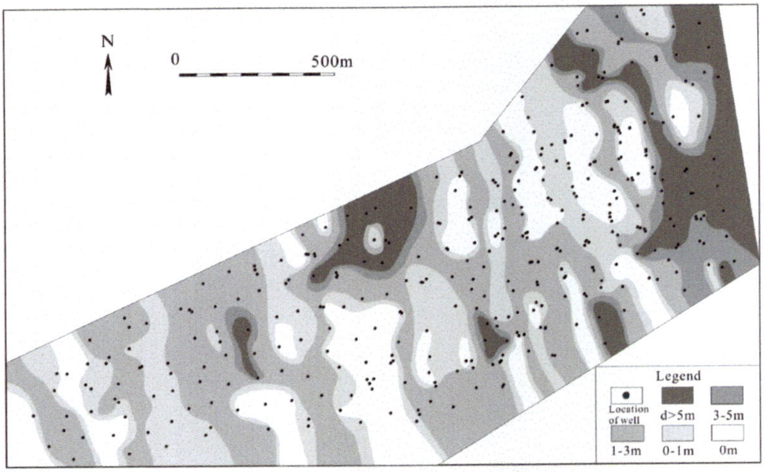

Fig. 5.24 Plane distribution map of interlayer development between yI35b and yI35c in a single layer of yI35b to yI35c in a certain area of western sag of Liaohe Basin

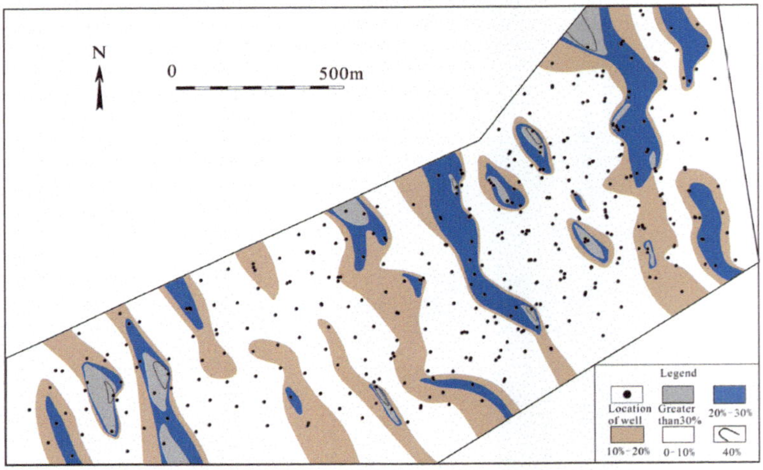

Fig. 5.25 Distribution characteristics of interlayer frequency in single layer yI11b in a certain area of western sag of Liaohe Basin

pathway, instead of isolating the steam. Qiu et al. (1996) set the thickness standard of interlayer under the secondary infill well pattern in the Daqing Oilfield as 1.5–2.0 m. According to the development experience of Liaohe Oilfield, the mudstone thicker than 1 m can effectively prevent steam migration. On this basis, the effective thickness of permeability barrier in the Yulou reservoir, where the permeability barriers are mainly mudstone and silty mudstone, is determined as 1 m. It is believed that the permeability barriers influence the steam flooding in the following aspects.

(1) Discontinuous mudstone increases the tortuosity of fluid flow pathways, complicating the injection–production relationships. The spatial distribution and extension of permeability barriers aggravates the reservoir heterogeneity (Fig. 5.26), making it challenging for the injected steam to advance uniformly in space and leading to the phenomena such as channeling and fingering, which reduce the displacement efficiency and complicate the pattern of fluid flow. This is similar to water flooding. It illustrates the plane distribution of permeability variation coefficient for yI_1^{1b}, where the red represents the permeability variation coefficient < 0.5, indicating weak reservoir heterogeneity; the gray represents mudstone; the yellow represents moderate heterogeneity; the blue represents the area with strong heterogeneity, which is dominant in the study area. Similar situations are found in the other 28 sublayers. Overall, the Yulou reservoir demonstrates strong heterogeneity.
(2) The numerous barriers/interlayers in the reservoir may absorb a large amount of heat in the steam flooding, weakening the effect of steam flooding; so, the economic feasibility, a critical indicator of steam flooding efficiency, should be considered (Liu et al. 1997; 1998a, b). Either barriers or interlayers are capable of absorbing the heat in steam flooding, but the former is much greater in absorbing capacity.
(3) During thermal recovery, the separate-layer steam injection is often used to expand the swept volume, prevent steam channeling, and ensure the uniform advancement of the steam front to improve the steam flooding efficiency. According to the plane distributions of the barriers and the steam injectors,

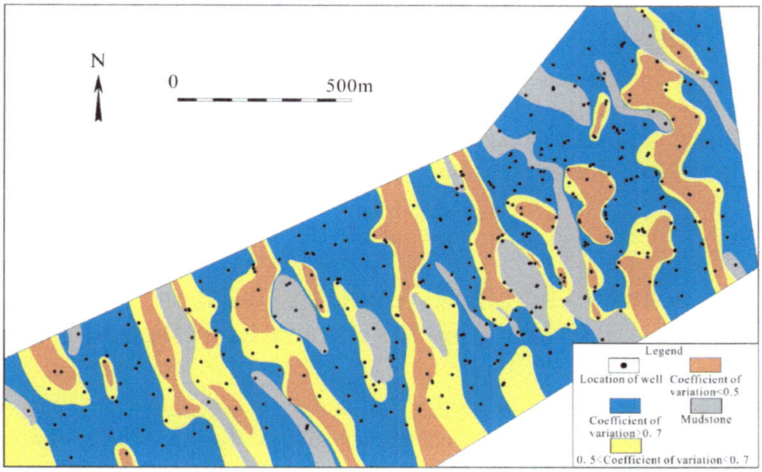

Fig. 5.26 Plane diagram of permeability variation coefficient in single layer of yI1 1b in a certain area of western sag of Liaohe Basin

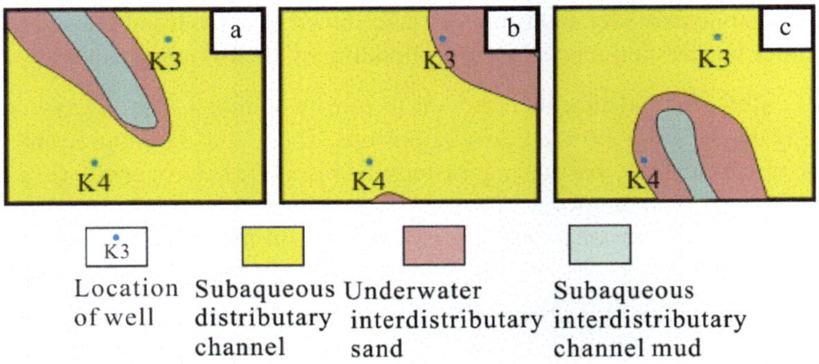

Fig. 5.27 Sedimentary microfacies distribution map of the main layers of K3 and K4 wells in a certain area of the western sag of Liaohe Basin

the separate-layer steam injection can be used in the regions with barriers to improve the development effect. To be specific, the sublayer-level sedimentary microfacies and barriers/interlayers are analyzed to clarify the effective thickness and plane distribution range of the thermally recovered heavy oil reservoir, and the layers to be thermally recovered are divided rationally. Then, the lower 1/2 or less of the steam injection layer is opened to mitigate the influence of steam overburden. Simultaneously, a steam foaming agent is used for interlayer or intralayer regulation, or the packer is used for separate-layer and ball-drop selective steam injections.

(4) The permeability barriers may result in poor injection–production relationship, impairing the steam flooding effect. Well K3 is a steam injector and Well K4 is an oil producer, spaced 107 m, in the northeastern part of the study area (Fig. 5.27). In the primary layers yI_1^2 ($yI_1^{2a} + yI_1^{2b} + yI_1^{2c}$; Fig. 5.27a), yI_2^3 ($yI_2^{3a} + yI_2^{3b} + yI_2^{3c}$; Fig. 5.27b), and yI_2^4 ($yI_2^{4a} + yI_2^{4b} + yI_2^{4c}$; Fig. 5.27c), underwater distributary interchannel sand or underwater distributary interchannel mud occurs between Wells K3 and K4, which act as the permeability barriers, abating the injection-production relationship. Well K3 reveals a five-year oil production of only 577 t, far lower than other producers in the same unit.

5.5 Sedimentary Microfacies Modeling Based on Geological Genetic Analysis in Fine Reservoir Characterization

Fine reservoir characterization refers to an integrated technology for comprehensively and precisely describing the geological characteristics and remaining oil distribution of the reservoir for rational assessment and development at the stage with high recovery rate and high water cut (Chen et al. 2017a, b). It has become a focus of studies in the mid- to late-stage of field development. A geologic model is central to fine reservoir characterization, and sedimentary microfacies modeling is one of the primary subjects of reservoir geological modeling research. Also, because the facies-controlled modeling (Yu et al. 2005a, b; Dong 2016; Duan et al. 2019) has become a common practice in reservoir geological modelling, sedimentary microfacies is prominently important. In a sense, the establishment of a 3D sedimentary microfacies model that best represents the actual reservoir conditions by combining various data and using multiple research methods/techniques (Ranie et al. 2006) is critical to the success of precise reservoir modeling and fine reservoir characterization.

5.5.1 Basis for Modeling

5.5.1.1 Sedimentological Analysis

Regional depositional setting should be understood before sedimentary facies analysis. Also, the sedimentary facies and subfacies are analyzed to build the depositional model, that is, the sedimentological analysis, thereby providing a solid foundation for sedimentary microfacies analysis. A research on sedimentary microfacies not based on facies and subfacies may yield "malposition of lithofacies". For sedimentary facies analysis, the geological methods include core observations and field outcrop surveys to directly capture sedimentary information recorded in subsurface reservoirs by analogy and analysis; the geochemical methods include various tests (widely applied in lithology, geological age and sediment source analysis), where proper geochemical indexes and methods are used to identify the sedimentary characteristics of reservoirs, considering that the contents of rock elements change with the geological processes, such as sedimentation and diagenesis; the geophysical methods are mainly logging and seismic data—logging and seismic facies is vital for sedimentary facies studies. These methods are often combined for corroboration and supplementation in practical studies to obtain more precise and detailed sedimentary information.

Moreover, the knowledge of sedimentary facies can be revised/added by flume simulations, computer-based depositional mechanism simulations, and numerical simulations. Some researchers have worked satisfactorily in this aspect (Mu 2000; Luo et al. 2003).

5.5.1.2 Establishment of Stratification Database by High-Resolution Sequence Stratigraphy

The stratigraphic division (into pay zones and layers) forms the basis for fine reservoir characterization (Guo et al. 2001), especially the sedimentary microfacies analysis. The sedimentary microfacies analysis in the mid- to late stage of development raises higher requirements for precision and genesis implications than that in the early stage. This is especially feasible with the support of high-resolution sequence stratigraphy. The primary tasks of high-resolution sequence stratigraphic studies during the fine reservoir characterization phase are the division and correlation of isochronous sedimentary stratigraphic units formed by high-frequency allogeneic cycles, aligning the reservoir architecture model and the isochronal stratigraphic framework with actual geology. The study precision can reach the layer or even the sandbody level. Layer division guided by high-resolution sequence stratigraphic theory has higher precision and more prominent genesis implications than traditional methods relying on thickness, cycle, and isoelevation slicing.

As implied by the name, the high-resolution sequence stratigraphic correlation technique is based on high-resolution sequence stratigraphy. It starts from genetic stratigraphic units and applies the depositional dynamic process–stratal response principle to understand the formation process and spatial distribution of strata. Researchers have different views on the degree of resolution required. I support that (1) the parasequences should be divided in thickness at the scale of tens of meters or even meters, by using a comprehensive sequence stratigraphic technique integrating core, log, outcrop, and seismic data, according to the classic sequence stratigraphic principle introduced by Vail; and (2) the parasequences should be divided to the level of layers in pay zone, in the high-resolution sequence stratigraphic study introduced by Cross (Wagoner 1990; Vail et al. 1991; Cross 1994; Zhu 2000; Yang et al. 2001; Zhao 2005).

5.5.2 Modeling Methods and Techniques

The methods/techniques are critical to the success of sedimentary microfacies modeling. As the researchers require increasingly capable models, the deterministic modeling technique has been rarely applied; instead, the stochastic modeling technique (Oriol et al. 2006) or its combination with deterministic modeling becomes popular. Two types of stochastic models are available: object-based and pixel-based (Hu et al. 2001), which are classified as the marked point process, sequential indicator simulation (Liu et al. 2003a, b), truncated Gaussian simulation, and multi-point geostatistics-based simulation. The multi-point geostatistics-based simulation has just emerged as a new technique, and its realization has not yet supported with any software. It prefers the data combination and multi-information coordination.

5.5.2.1 Three-Dimensional Sedimentary Microfacies Modeling Based on Horizontal Well Data and Outcrop Information

Outcrop survey has received much attention from geologists over the past years as a major method for fine reservoir characterization and modeling. It is gradually acknowledged that only the investigation on outcrop reservoirs with the same sedimentary types as the target of interest can yield more detailed and accurate geologic knowledge and reservoir prediction techniques than dense well patterns (Qiu et al. 2000; Wu et al. 2002a, b; Matthew et al. 2007). Horizontal well data and outcrop information can be applied to effectively extract the geostatistical parameters (especially those related to the lateral variability of sandbodies) required by the three-dimensional modeling. Wu et al. (2003a, b) applied a single sandbody width of 40–100 m determined from horizontal well data and outcrop information to the three-dimensional sedimentary microfacies modeling of the Triassic Yanchang Formation in the Pingqiao region, the Ordos Basin, with good practical results. In three-dimensional sedimentary microfacies modeling, it is challenging to predict the interwell connectivity and continuity of sandbodies. Most development wells in oilfields are spaced at hundred or hundreds of meters. However, the single sandbody width is usually less than that the well spacing, and the seismic resolution is insufficient to recognize the single sandbody less than 5 m thick. These bring challenges to interwell sandbody prediction. The researchers have attempted to optimize the three-dimensional reservoir modeling by using the geostatistical stochastic modeling method, with the premise that the geostatistical characteristics of the modeling area must be understood in advance; however, for the modeling area where the sandbody width is less the well spacing, the sandbody's statistical characteristics are limited by well spacing. This issue may be addressed by introducing horizontal well data, which are inherently a direct representation of the sedimentary sandbody distribution, with outcrop data collected from adjacent areas into three-dimensional modeling. However, this method is limited for application, because the horizontal well data and outcrop information are often rarely enough.

5.5.2.2 Three-Dimensional Sedimentary Microfacies Stochastic Modeling Based on Geologic and Seismic Data

Zhou et al. (2002) used well log data to constrain seismic data in studying the Z oil reservoir in the F Oilfield, Venezuela, to build a three-dimensional seismic impedance model with high vertical resolution, and adopted the stochastic inversion of 3D seismic data volume and well data to constrain the stochastic facies modeling to reduce the uncertainty. The results correlate well with the geologic, well log, seismic, dynamic, and static data.

Wu et al. (2003a, b) used impedance information and stochastic modeling to study the fluvial reservoir in the Neogene Minghuazhen Formation in a block of the Bohai Bay Basin. Central to the method is using well data and seismic data to predict the sedimentary microfacies in a three-dimensional space. Any seismic-based prediction of the reservoir's lateral extent is a transition from seismic attributes to geological parameters. Given the nonuniqueness inherent to seismic attributes, this transition is often unsatisfactory. Stochastic simulation admits no strict deterministic relationship between impedance and geological parameters. Instead, it attempts to find the probability correlation between them and uses this correlation to perform stochastic modeling of reservoirs. The uncertainty associated with prediction is assessed to solve the problem of nonuniqueness. This method can obtain a three-dimensional impedance data volume by running the log-constrained inversion. The next step is extracting impedance data at near-well traces. The probability correlation between the well model of sedimentary microfacies (e.g., channel sandstone, overbank sandstone, and floodplain mudstone) and the impedance at near-well traces is analyzed to estimate the probability at each impedance value to indicate different sedimentary microfacies to achieve the objective of sedimentary microfacies modeling.

5.5.2.3 Stochastic Simulation Modeling Based on the Marked Point Process

Hu et al. (2002) applied the stochastic simulation modeling based on the marked point process to the quantitative characterization and three-dimensional modeling of sedimentary microfacies in low-permeability reservoirs. The object-based method (with an object as the basic simulation unit) applies the marked point process model and optimized algorithm (e.g., simulated annealing or Metropolis–Hasting) to conduct the stochastic simulation of discrete objects. The Poisson point process assumes that the central points of objects are spatially independent (i.e., the probability distribution theory of the Boolean model), whereas the Gibbs point process assumes that they are mutually related or exclusive. When the objects' positions are independent, and the object density (the average quantity of object per unit reservoir volume) is constant, the position of objects' central points conforms to the stationary Poisson point process with the premise that marked points are random. This method applies to the sandstone intercalated with small-scale mudstone or the mudstone containing small-scale isolated sandstone. When the positions of objects are not independent and related (superimposed), it becomes the Gibbs point process. This method applies to stacked channels in a fluvial sandstone belt, such as fluvial or fluvial delta settings and associated sedimentary facies. This modeling method has two advantages. First, it is simple and flexible, with which some newly acquired and calibrated geological data can be readily introduced into the model as conditional information to maximize the knowledge from geologists to realize a human–computer interactive modeling process. Second, it is a valuable tool for conducting quantitative characterization and three-dimensional modeling of sedimentary microfacies. This method provides a series of equal probability models. These models overcome the uncertainty associated with sedimentary facies boundaries in regions with no well control. Also, they conform to well and seismic data and provide a more objective representation of sand size, continuity, and contact mode. Thus, this method is suitable for modeling the distribution of heterogeneous, complex sedimentary microfacies, reflecting the variation rule with a high reliability.

5.5.2.4 Modeling Based on Geological/Geophysical Log and Development Dynamic Data in Combination with Geological Knowledge

I modeled the stratigraphic–lithologic reservoir of the Triassic Yanchang Formation in a block of the Ordos Basin in 2006. The study area is covered by loess tableland. Due to a lack of seismic data, various data (e.g. outcrop, core, reinterpreted log, and development dynamic data) are adsorbed, and the geological knowledge of researchers is integrated. Since the interval of interest comprises fluvial deposits, the RMS software and the pixeled-based modeling method are selected. In the process of modeling, a detailed analysis is made on log microfacies (interpreted log microfacies are assigned to the 0.125 m thick interval of interest in each well). A variogram is fitted manually by the statistical analysis of geological data (e.g., regional geology, core, and outcrop), geophysical log data, and development dynamic data to determine the major, minor, and vertical ranges of the 3D microfacies model in space. When building each model, independent analysis is conducted for each type of sedimentary microfacies in every layer, with information from the statistical analysis of outcrop and core data. Then, boundary constraints and other measures are used to build a stochastic model to characterize the three-dimensional distribution of sedimentary microfacies in the reservoir. It shows the workflow (Fig. 5.28).

The pixel-based method (with pixel as the primary modeling unit) combines the pixel-based stochastic model and various algorithms. For instance, the sequential simulation algorithm can be applied to the Gaussian model to form the sequential Gaussian simulation, or applied to the indicator simulation to form the sequential indicator simulation. The first step in building a sedimentary microfacies model is the variogram analysis to specify the major, minor, and vertical ranges, thereby allowing the accurate prediction of the three-dimensional characterization of sandbodies. A variogram describes the structural changes of a regionalized variable and illustrates its random variation. Furthermore, the variogram calculation forms the basis for other geostatistical calculations. A variogram is a crucial tool for characterizing the random function and measuring the spatial variability of a regionalized variable. It reflects how the spatial variation characteristics change with distance. The parameters (e.g. range, nugget, and sill) illustrated in the variation diagram (the variation of a variogram with lag distance) can represent these characteristics. In the previous study, the thoughts discussed above are realized using RMS, a representative software developed by ROXAR. This software retains the advantage of Storm in sedimentary facies simulation, and its essence is the stochastic simulation method of reservoir sedimentary microfacies established based on the sedimentary reservoir system and genetic unit theory. The final sedimentary microfacies model is validated satisfactorily using several methods. This method is capable of precisely analyzing the logging microfacies, validating the development dynamic data, and realizing the human–computer interaction. However, it is disadvantageous for heavy workloads. For instance, significant efforts must be spent on log facies analysis for the entire wellbore and manual variogram fitting.

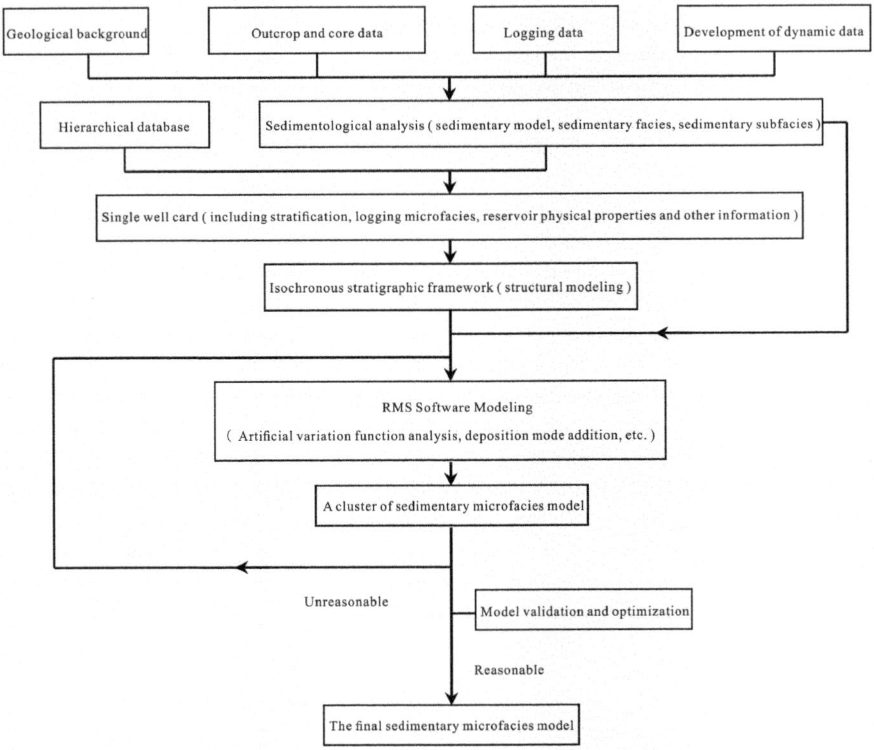

Fig. 5.28 The flow chart of sedimentary microfacies model is established by combining logging data and oilfield development dynamic data with geological thinking

5.5.2.5 Modeling Based on Architectural Analysis

An architectural analysis must be conducted to improve the reservoir depositional model's precision. In the analysis, a hierarchy of eight orders of architecture boundary (Fig. 5.29; Miall 1985, 1988, 1996) with varying durations is defined to separate the sedimentary units at various scales. The first-order boundary is the interface of the cross-bed sets, at which rare or no erosion occurs. This boundary is invisible on cores, and generally can be identified according to the truncation and pinch-out of cross-bedded forest beds. The second-order boundary is the interface of pay zones, showing lithologic contrasts. The third-order sequence is the erosion surface with a low dip angle (usually less than 15°) inside the giant bedform. It truncates the underlying cross-bed set(s) and is typically draped with a set of mudstones overlain by intra-clastic mud pebbles. It is frequently associated with ambiguous lithologic contrasts. The fourth-order boundary is the interface of the giant bedform, such as the interfaces of channel bar, point bar, and crevasse splay, as well as the scouring surface at the base of small-scale channels (overflow channels). The fifth-order boundary delineates the interface of large-scale sand sheets, such as large-scale channels and channel fill complexes. It frequently shows a flat top and concave upward geometry and is marked by incision–fill landforms and lag gravel deposits. The sixth-order boundary is the interface defining the extent of a channel group or incised valley, roughly equivalent to the interface of the member or submember (the stratigraphic unit for mapping). The seventh-order boundary delineates the sedimentary bodies associated with allogeneic cycle events, such as the maximum marine (lacustrine) flooding surface. The eighth-order boundary is the regional unconformity delineating the third-order sequence (Miall 1985, 1988; Hu et al. 2002). The objective of defining the architecture boundary is to use a set of hierarchal bedding contacts to separate a sandbody into genetically-related stratal packages. Architectural analysis at different scales can improve the understanding of the depositional characteristics of reservoirs being studied (Joseph et al. 2006; Steven et al. 2006; Miall 2006) and precisely divide the sedimentary facies into various microfacies to build a sedimentary microfacies model close to the actual subsurface geology. Architectures are analyzed by several methods/techniques, such as core, outcrop, and ground-penetrating radar (Lee et al. 2007). Architectural analysis has become a frontier domain for sedimentary microfacies studies and is increasingly crucial in sedimentary microfacies modeling.

However, the application of architectural analysis is limited to certain extent. It is applicable to the sedimentary strata with the alluvial action being the dominant transport mechanism, but not to the circumstances such as gradual lithologic transition, single surface type, ambiguous boundary order, intensive diagenetic reformation, synsedimentary deformation, and strong diapirism. Thus, the architectural analysis

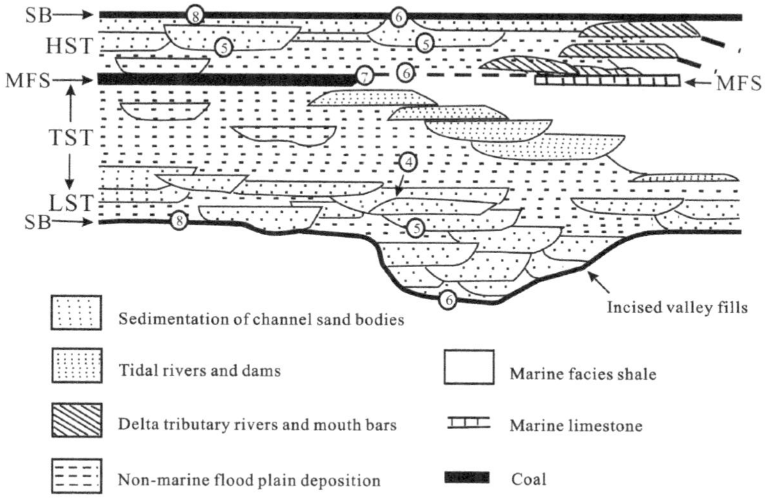

Fig. 5.29 The schematic diagram of reservoir architecture interface at all levels (according to Miall 1996)

method is not recommended when dealing with sedimentary bodies deposited in a river-mouth bar, intertidal zone, and fluvial delta front settings (Lan et al. 2001).

5.5.2.6 Modeling Based on Cross-Well Seismic Technique

The cross-well seismic technique was introduced in the 1970s into the petroleum exploration and development. Seismic waves generated by a source placed in one well are recorded by receivers placed in another or other wells. Compared to surface seismic, the cross-well seismic technique is characterized by shorter distances of energy transmission, proximity to the detection target, and the ability to avoid the low-velocity zone. Thus, the data recorded have high frequencies and signal-to-noise ratio (SNR). Combining cross-well seismic survey and other techniques (e.g., well logging) integrates stochastic and deterministic modeling. This method improves the precision of interwell prediction and permits fine characterization of interwell sedimentary microfacies. The cross-well seismic technique uses two types of information: (1) formation velocity acquired from the first break travel-time tomography, and (2) reflection information acquired from reflection wave field imaging. The former is usually used for interwell lithological analysis and fluid prediction, and the latter is a powerful tool for detailed studies of interwell reservoirs (Cao et al. 2006).

The cross-well seismic technique is uniquely capable of (1) performing high-resolution monitoring of interwell reservoirs, (2) yielding a resolution 10 to 100 times of surface seismic, (3) generating two-, three-, and four-dimensional interwell reservoir information, and (4) providing the depth-domain data, which can be correlated directly with well log data, thereby bridging the well and seismic data (Chen et al. 2003). When the cross-well seismic technique is applied to solve the issues associated with describing oil reservoir properties, the precise cross-well seismic inversion can provide a more accurate representation of the reservoir geometry and physical properties to optimize the heterogeneous oil reservoir model; thus, it helps improve the precision of reservoir characterization, providing an adequate support for interwell prediction in sedimentary microfacies modeling. The cross-well seismic data exhibit high frequency and SNR, and they can be used to detect thinner oil zones and more subtle geological characteristics that cannot be achieved with conventional surface seismic surveys or VSP. Examples have shown that the cross-well seismic technique extends the seismic image with a level of resolution close to that of the well log to the interwell domain (Cao 2002). Thus, it is applied to field development for detailed reservoir characterization and dynamic oil and gas monitoring. The cross-well seismic technique is advancing toward practicability and has become an effective tool for fine reservoir characterization, dynamic reservoir monitoring, and other studies. The technique is restricted by high acquisition cost and limited detection space; however, these issues will be addressed with advancements in the cross-well seismic technique.

5.5.3 Key Considerations in Modeling

5.5.3.1 Selection of Modeling Method/Technique

Various methods/techniques are available for studying sedimentary microfacies. They are summarized as follows.

(1) Basic geological methods/techniques

For instance, core and outcrop data are used to study sedimentary microfacies, provide the most direct representation of geological conditions, and have unreplaceable advantages compared to any other method. Horizontal well data combined with outcrop information for sedimentary microfacies modeling mentioned in the previous discussion is a typical example. Ongoing research on architectural analysis methods is popular domains in geological methods that serve as tools for studying sedimentary microfacies. Major challenges associated with this method include the data paucity, high workload, and limited applicability.

(2) Computer, mathematical geostatistical, and mathematical methods/techniques

These methods/techniques run the gamut from general geostatistical methods to modeling software developed by combining computer technology with mathematics. Geostatistics is based on the theory of regionalized variables and uses a variogram as the primary tool to statistically analyze random and structural variables. Geostatistical methods have played a vital role in China's sedimentary microfacies studies. To meet the requirements of production practice requires a more in-depth understanding of geostatistical theory and a broader range of methods and principles. Besides the most studied ordinary kriging methods, more in-depth studies on various methods were conducted, such as nonstationary linear geostatistics, nonparametric geostatistics, multivariable geostatistics, and spatiotemporal multivariable geostatistics, which remain a frontier domain of the discipline (Qi et al. 2007). Substantial breakthroughs have been made in the application of these methods/techniques, with several software systems developed for domestic production (Hou et al. 2008). Moreover, modeling software is continually updated with advancements in computer and related disciplines to embody geostatistical ideas. Some software for deterministic modeling includes SGM, EarthVision, and Geofram. Software applicable to stochastic modeling includes RMS, GSLIB, Herisim, RC2, GOCAD, and GridState (Deutsch et al. 1996; Wu et al. 1999). Note that mathematical statistical methods will be a game of numbers without considering the geological meaning. Also, when developing and using various modeling software, it is required to select the modeling software applicable to the depositional setting and closely relate modeling principles to geology. Otherwise, the model can only produce a visual effect, but cannot reflect the actual features of subsurface sedimentary microfacies, and fail to produce an effective prediction.

(3) Geophysical methods/techniques

The most common data used are well log and seismic data. The former includes spontaneous potential, natural gamma-ray, resistivity, and formation dip, and emerging technologies, such as near-borehole acoustic and resistivity imaging (Yu et al. 2005a, b). The latter includes seismic impedance-constrained inversion, four-dimensional seismic, cross-well seismic, prestack seismic inversion, and multiwave and multicomponent seismic (Wang et al. 2006; Liu et al. 2012).

Different modeling methods/techniques and data processing modes have distinct features and are applied to various sedimentary facies. The differences in depositional setting and data availability limit the selection of modeling methods/techniques to some extent. Thus, to find an appropriate modeling method/technique, various data will be combined to accurately understand the modeling object and requirements (Feng et al. 2016). The methods/techniques can be organically integrated to yield a precise interwell prediction.

5.5.3.2 Selection of Modeling Software

Three-dimensional geologic models of sedimentary microfacies and the geological knowledge of each algorithm rely on modeling software. Modeling will likely be unsuccessful unless a suitable software is selected. The appropriate modeling software should (1) be capable of accurately predicting interwell sedimentary microfacies, (2) be well-suited to the study area regarding depositional setting and model, (3) incorporate the researchers' geological knowledge as much as possible, (4) reflect the latest knowledge and technologies associated with the geological, geophysical, computer, and mathematical realms, and (5) meet the requirements of a large-scale application with improved precision and predictability and reduced workload.

5.5.3.3 Principles for Facies Model Validation and Selection

The result of microfacies modeling is the sedimentary microfacies model, which provides an essential tool for predicting subsurface sediments. The microfacies model validation and selection are vital to the success of the subsequent modeling of reservoir physical properties (for facies-controlled modeling especially). Therefore, assessing whether the facies model is reasonable and geologically realistic is critical (Zhou et al. 2001). The principles for microfacies model validation and selection are summarized as follows.

(1) The model contains geological meaning and honor geological concepts. In some sense, stochastic modeling represents geological phenomena numerically. Hence, the parameters should reproduce the features of real subsurface geobodies, such as the size, thickness, extending direction, and shape.

(2) The model is consistent with the manually mapped sedimentary microfacies. This practice is traditional and direct and uses the researchers' geological knowledge to build the sedimentary microfacies model and correlates it with the software-built model. The higher the degree of consistency is, the better the modeling quality is.
(3) The realization of rarefying the well is close to realizing a dense well pattern. Several realizations of the conditional simulation are generated by randomly rarefying the well pattern. It is required to record the facies types and probability distribution of sand thickness at each rarefied well location and observe the changes in the sandbody's shape, size, and distribution. A significant characteristic of stochastic modeling is the random distribution, which represents the uncertainty associated with subsurface geology. Thus, after rarefying, well differences between realizations will exist. The model's precision can be validated in the allowable error range by assessing the consistency between the models before and after rarefying.
(4) The modeled sedimentary microfacies are validated using the development dynamic data. If a model comprising an oil producer and a water injector nearby predicts that the channel deposits at two wells are connected, but in the practical development, no response is obtained in the oil producer after water injection, then, the accuracy of the sedimentary microfacies model should be questioned. When modeling for the Yanchang Formation in an area of the Ordos Basin, I found that the perforated oil-bearing intervals are mostly channel deposits. I applied the perforation data to verify the modeling of the main channel deposits, and also used the development dynamic data to validate the geological model, suggesting reliability and feasibility.

5.5.4 Problems and Prospects of Sedimentary Microfacies Modeling

5.5.4.1 Problems

As discussed above, the methods/techniques for modeling sedimentary microfacies, including basic geological method, geostatistical method, stochastic modeling that integrates computer techniques with mathematical methods, and geophysical methods (e.g., well logging and seismic), have their respective merits and defects. Some defects are caused by the degree of research and the actual geological conditions, whereas others by the limited level of current scientific and technological development. In reservoir characterization, the researchers should consider carefully to choose the strategy and methods/techniques of modeling suitable for the research object, aiming to avoid or even eliminate the defects (Yin et al. 2006). When studying sedimentary microfacies, the researchers should firmly follow the prediction of interwell microfacies at a meter scale. On one hand, the modeling technique, method, and software applicable to the study area are selected on the basis of sedimentary knowledge of the study area. On the other hand, the researcher's geological knowledge should be incorporated into the modeling process to improve the modeling precision and the geological meaning of algorithms.

5.5.4.2 Prospects

Three-dimensional sedimentary microfacies modeling uses multidiscipline information (Yin et al. 2007) to characterize the distribution of sedimentary microfacies in three-dimensional space. It is a frontier in reservoir geologic research and characterization, and has critical practical significance in improving the benefit of field development (Wu et al. 2002a, b; Chen et al. 2006a, b). In the future, the sedimentary microfacies modeling is believed to develop in the following prospects.

(1) Prototype models of various reservoir types and geological knowledge bases are established and improved on the basis of detailed field and subsurface geology studies to provide reliable sedimentary conceptual control and geostatistical guidance to sedimentary microfacies modeling.
(2) Attempts are made to explore more optimized algorithms, develop new modeling software based on multivariable geostatistical methods (by applying training image, that is, the reservoir geological mode), and incorporate the researcher's geological knowledge into the modeling process.
(3) Various modeling methods/techniques are combined as much as possible to improve the modeling precision and predictability.
(4) The latest technologies and research results of multiple disciplines, such as geology, geochemistry, and geophysics, are incorporated into the modeling (Yin et al. 2017) to improve the modeling precision and efficiency.
(5) Stochastic modeling is integrated with deterministic modeling to ensure high precision and predictability.
(6) Considering the features of dense well patterns at the mid- to late-stage of development, the development dynamic data are added to the sedimentary microfacies modeling to improve the precision, predictability, and practicability of the model in application to field development.

5.6 Classification and Evaluation of Geobody Genesis in Thermally Recovered Heavy Oil Reservoirs

A geobody functions as the basic unit of steam flooding and it is the material basis for thermal recovery. An in-depth and proper understanding of geobodies directly influences the effectiveness of steam flooding. As a conventional practice, a heavy oil reservoir is thermally recovered by steam huff-and-puff and then by steam flooding. In the stage of steam huff-and-puff, geobody studies are considered less critical because the drainage area is limited to a small extent around the producing well. In the stage of steam flooding, geobody studies become necessary, because the drainage area related to thermal recovery increases significantly due to the connectivity between the steam injector and the oil producer, coupling with the fact that the reservoir is highly heterogeneous. The thermal recovery of heavy oil has specific requirements on the effective thicknesses (generally 10 m) of pay zones to be produced. In this book, the term "geobody" refers to distinct geologic units that spatially contact in a set of thermally recovered pay zones (e.g., yI and yII), which produce heavy oil by steam flooding. It comprises multiple sandbodies vertically separated by barriers/interlayers and laterally extending to a specific distance. The geobodies are combined deposits formed at a specific geologic time and roughly correspond to the third-order sedimentary cycle (or sequence). Extensive research has been conducted on geobodies related to petroleum exploration and development and various solid minerals (Chi et al. 1999; Zhao et al. 2010). Du et al. (2004) classified the geobodies into the reservoir-, formation-, intrastratal-, and pore-space scales, in order from macroscopic to microscopic, and analyzed the formation and distribution of remaining oils in geobodies of different scales. Wang et al. (2008) presented the modeling and 3D visualization of geobodies. Zhai et al. (2009) discussed the relationship between block petroliferous geobodies and petroleum exploration. Wen et al. (2011) applied curvature to detect complex geobodies and suggested that the fault and channel are readily recognized by combining the maximum positive curvature and the minimum negative curvature. Liu et al. (2012) studied the hydrocarbon transport system of a complex geobody in the Chepaizi region at the Junggar Basin's western margin. Wu et al. (2013) discussed an architectural hierarchy scheme of clastic sedimentary geobodies. Fan et al. (2019) introduced the approach of storage and reconstruction of three-dimensional geobody models, which has been proven effective in reducing the model's storage volume and strengthening the data migration, sharing, and exchange. He et al. (2020) further refined the three-dimensional entity modeling approach to a complex geobody and illustrated the specific workflow for modeling four types of complex geobodies. Zheng et al. (2020) explored the implication and evaluation technology of the integrity of a geobody as reservoir-type underground gas storage.

The existing studies mostly deal with geobodies related to waterflooded reservoirs and solid minerals, but few efforts have been conducted on classification of geobodies in thermally recovered heavy oil reservoirs. This section presents a study on the geological genesis of geobodies and conducts classification and evaluation of geobodies based on characteristic parameters representative of their basic properties, with an aim to provide a reference for the steam huff-and-puff to steam flooding conversion. Based on an overview of previous studies regarding the geological factors influencing steam-flooded reservoirs, the parameters that can adequately represent the factors influencing the characteristics of reservoir geobodies are selected, such as effective thickness, net-gross ratio (NGR), porosity, permeability, permeability variation coefficient, and interbed frequency. These parameters are used to classify and evaluate the geobodies in thermally recovered heavy oil reservoirs. The information on the study area and the target horizons are detailed in Sect. 5.2, Chapter 3.

5.6.1 Major Geological Factors Influencing the Steam Flooding

Steam flooding for thermally recovered heavy oil reservoirs differs significantly from conventional water flooding. For instance, the steam flooding suffers from steam energy consumption by barriers/interlayers and steam overlap, which will never happen in water flooding. Therefore, it is considered that steam flooding can be influenced by more factors and has more complex processes than water flooding. Many scholars have been studied steam flooding through physical and numerical simulations. Liu (1997) divided the primary reservoir geological conditions and fluid properties influencing the effectiveness of thermal recovery into five parameter groups through simulation and other methods: (1) crude oil viscosity and relative density, (2) reservoir depth, (3) net thickness and NGR, (4) porosity, original oil saturation, and reserve coefficients, and (5) reservoir permeability. Physical simulation tests, numerical simulation, and oil field practice have demonstrated the influence of these parameters on the thermal recovery of heavy oil. Based on previous studies, the available data of the Yulou reservoir in the Liaohe Basin's West Sag are correlated. The viscosity and density of crude oil in the northwestern part of the study area are higher than that in the southeastern part. Therefore, steam flooding effect should be better in the northwestern part than in the southeastern part. The bottom of the Yulou reservoir is not deeper than 1200 m, which is not a problem in a steam flooding operation because it can be applied to oil reservoirs buried at depth up to 1600 m (Liu 1997). An essential condition required by steam flooding on an oil zone is that the oil zone should be sufficiently thick and contain as few as possible tight interlayers (e.g., shale, argillaceous siltstone, and clay bed). A thicker oil zone is preferred because the heat energy that flows into it will continually be lost toward the roof and floor. Furthermore, attention should be paid to five geological characteristics (Liu et al. 1997): (1) reservoir lithology, (2) oil zone pressure, (3) low formation dip, (4) good connectivity between injector and producer, and (5) the interference of edge and bottom water. The Yulou reservoir is mainly composed of inequigranular sandstone and medium- to fine-sandstone, lithologically suitable for steam flooding. Because the steam huff-and-puff method has been implemented in the study area and decreased the formation pressure compared to the original pressure, the current formation pressure is suitable for steam flooding. The formation dip ranges from 2° to 10°, which will not cause an intensive overlap of steam. The connectivity between injector and producer is also a key concern. The architectural analysis of sandbody reveals good connectivity between injector and producer within the same underwater distributary channel sandbody. Thus, a good steam flooding considers the architectural analysis of sandbody. The formation pressure should be appropriately managed to mitigate the risk of a rapid and sharp decrease, given the edge and bottom water. Also, the preferred pathway for water channeling should be isolated with a packer. The term "sandbody" here refers to a sandbody that is aerially and vertically continuous but separated by mudstone, or other impermeable interlayers from the underlying and overlying sand (Zhang et al. 2008a, b). The sandbodies may exist in distinct geobodies. In this study, geobodies in yI and yII are divided individually. Thus, a geobody can contain multiple single sandbodies.

Based on the analysis above, the parameters such as effective thickness, NGR, porosity, permeability, permeability variation coefficient, and interbed frequency are selected as the standard parameters for classification and evaluation of geobodies. Effective thickness and NGR are the primary parameters, porosity and permeability are the secondary parameters, and permeability variation coefficient and interbed frequency are the reference indexes (Table 5.2). Using the geostatistical method and supported by experiences concerning the reservoir flow unit and pore structure (Chen et al., 2011a, b, 2013a, b, c), the geobodies in the Yulou reservoir are classified and evaluated. To ensure the weights of parameters and the correctness of classification results, the following principles must be followed:

(1) The result of each class should not be less than 5% of the total data points to locate and remove the outliers, if any, thereby eliminating the impact of outliers.
(2) There should be no apparent contradiction between the classification results and the current development performance of the oil field (e.g., steam injection volume, liquid rate, and oil rate from steam huff-and-puff wells). Care must be taken to not to spend an effort on waterflooded wells because rapid water flooding will result in an exceptionally low oil rate even from Class A geobodies.
(3) The primary parameters (effective thickness and NGR) and secondary parameters (porosity and permeability) should be fully considered in the classification. When an inconsistency occurs, the primary parameters shall prevail and be used to adjust the weight to the evaluation parameters and the classification and evaluation results.
(4) The classification results should be consistent with the geological analysis results. For example, the fan delta front subfacies in the Yulou reservoir can be divided into underwater distributary channel, underwater distributary interchannel sand, river-mouth bar, front sheet sand, and underwater distributary interchannel mud, with the underwater distributary channel, underwater distributary interchannel sand, and river mouth bar in dominance. Regarding sand size and thickness, Class A geobodies typically correspond to underwater distributary channel, while Class B geobodies mostly correspond to underwater distributary interchannel sand. In the Yulou reservoir, diagenesis is not influential, so sedimentary microfacies provide the most vital evidence among all geological factors. The correspondence between the sedimentary microfacies discussed here and the geobody is a concept of dominant facies because a reservoir must be sufficiently thick to be thermally recovered. A geobody may contain several single sandbodies of distinct sedimentary microfacies (Fig. 5.30). Geobody classification and evaluation are realized by repeating the test, adjustment, revision, and improvement, subject to the four principles above. The geostatistical analysis is completed using Excel (Fig. 5.31), and parameter weights are adjusted manually.

Note that the geobody classification and evaluation is not a simple and mechanical repetition process. Instead, geological factors are introduced in each test round

Table 5.2 Table of geological body classification results of Yulou oil layer in a certain area of western sag of Liaohe Basin

Parameter classifictaion		Effective thickness (m)	Net total thickness ratio	Porosity (%)	Permeability ($\times 10^{-3}$ μm^2)	coefficient of variation/Vk	Sandwich frequency (layer/meter)
TypeA	Maximum	56.57	0.662	37.2	3895.25	6.95	0.354
	Minimum	30.11	0.343	29.89	432.89	0.00	0.126
	Average	37.43	0.457	33.23	1796.68	3.34	0.207
Type B	Maximum	46.61	0.548	36.75	4090.41	6.71	0.309
	Minimum	12.29	0.184	27.43	131.55	0.00	0.06
	Average	25.76	0.336	32.59	1624.66	3.13	0.177
Type C	Maximum	47.22	0.562	38.45	5179.25	7.45	0.557
	Minimum	0.00	0.000	0.00	0.00	0.00	0.026
	Average	4.65	0.070	29.43	423.76	2.81	0.124

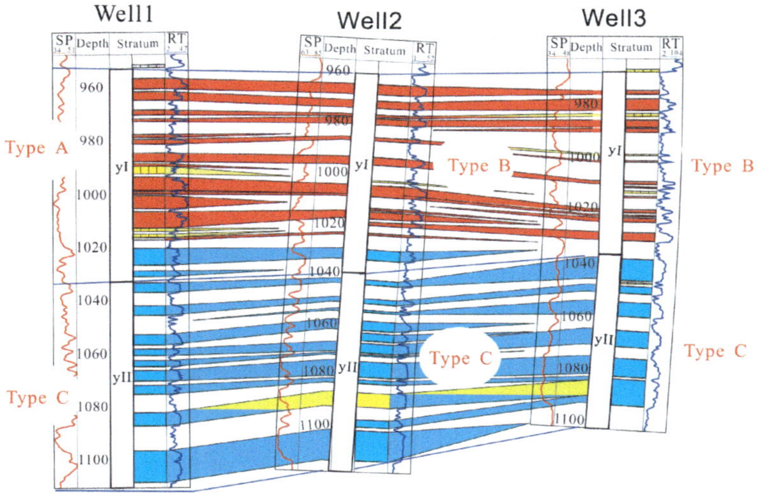

Fig. 5.30 Classification characteristics of single well geological bodies in Yulou oil layer in a certain area of western sag of Liaohe Basin. *Note* Legend same with Fig. 5.32

to constrain the geostatistics with researchers' geological knowledge. It is a real human–computer interaction process that enables a close integration of qualitative geological analysis with quantitative statistics and can produce the optimum geobody classification and evaluation (Chen et al. 2015a, b).

5.6.2 Classification and Evaluation of Single-Well Geobodies

As mentioned earlier in this book, the Yulou reservoir is divided into two pay zones: yI and yII. Considering the current development conditions, geobody classification and evaluation are conducted for yI and yII, separately. Specifically, the geobodies in yI and yII are classified and evaluated from the perspective of effective thickness, NGR, porosity, permeability, permeability variation coefficient, and interbed frequency, after the geostatistical analysis of 470 valid data points. The geobody classification on a single-well basis illustrates that Classes A and B geobodies are present in yI, whereas Class C geobody dominates yII. Therefore, yI should be the primary target for the steam huff-and-puff to steam injection conversion. As illustrated, Class A geobodies contain thick oil zones (> 35 m), with high NGR (up to 0.45), porosity of over 30%, and permeability mostly > 1500 mD (at least 500 mD). Class C geobodies contain oil zones generally thinner than 5 m, with NGR < 0.1, porosity < 30%, and permeability < 400 mD. Class B geobodies lie between Class A and Class C regarding the properties discussed above and is defined as the secondary target for steam flooding. Class C geobodies are predominately mud zone or water

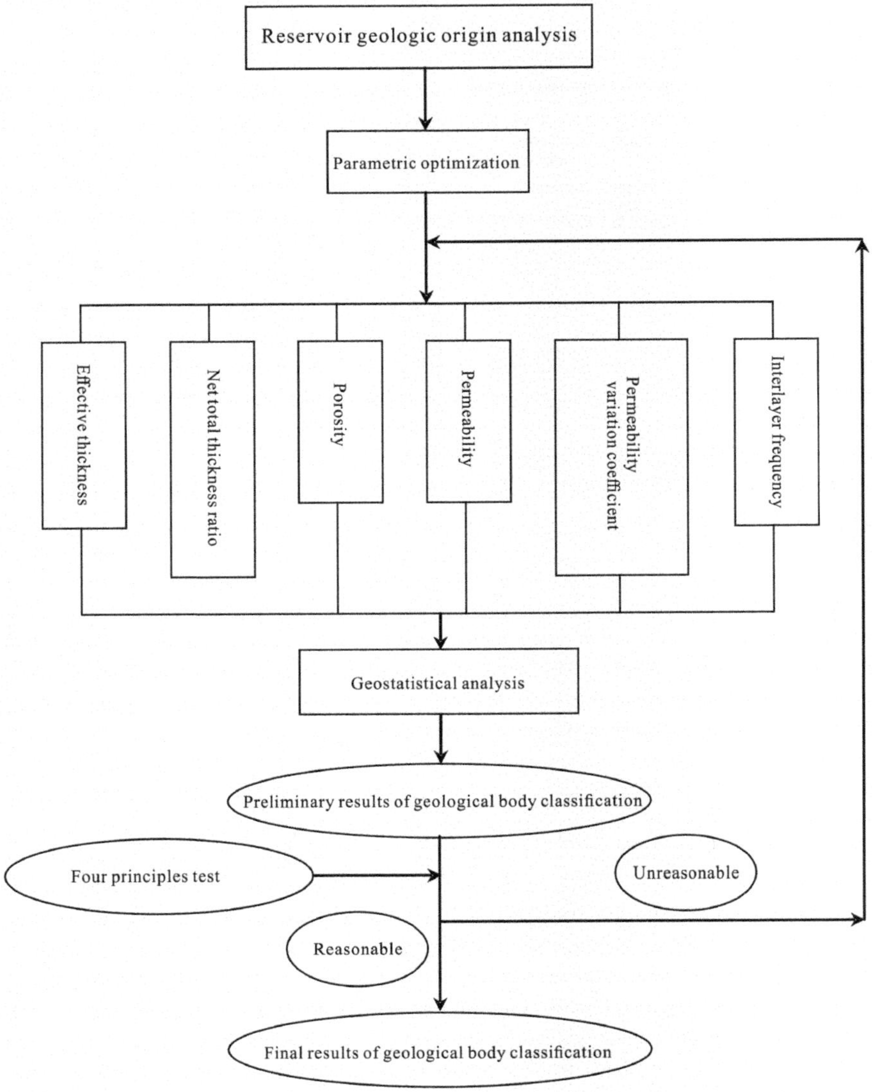

Fig. 5.31 Geological body classification flow chart of Yulou oil layer in a certain area of western sag of Liaohe Basin

zone, and economically unviable for development. From the perspective of geological genesis, Classes A and C geobodies mostly correspond to underwater distributary channel sandbodies. However, the former shows the dominance of oil zone, whereas the latter shows the dominance of water zone. These sandbodies contain thick sand layers, with a great lateral extent and good physical properties (e.g., high porosity and high permeability), with a loose recovered core. Class B geobodies correspond to underwater distributary interchannel sand or sheet sand microfacies. These sand-bodies are thin with poor lateral connectivity, short extension, and low porosity and permeability.

5.6.3 Classification and Evaluation of Flooding-Unit Geobodies

A flooding unit is the most basic development unit. Classification of geobodies in a flooding unit or different flooding units can provide direct guidance for steam flooding. Here, the cross-well profile is plotted from the geobody classification for wells. With this profile, the geobody type changes in a flooding unit are analyzed to define the interwell reservoir connectivity and thus assess the effect of steam flooding. A geobody has considerable scale in space. Single-well geobodies reflect the information at a specific well location, indicating the geobody type penetrated by a well. Flooding-unit geobodies, however, cover a specific spatial extent. A flooding unit can contain one or more geobodies of the same or different types—given the same types, the reservoir connectivity within the flooding unit is good (for Classes A and B geobodies); given the different types, the reservoir connectivity within the flooding unit is poor. Either single-well or flooding-unit geobody is just a component of the study process. A geobody has only one true concept, as introduced in the Preface.

Furthermore, the geobody classification and evaluation of different flooding units is correlated with the current production status of the oilfield. It shows the correlation between the geobody classification and evaluation and liquid production for the flooding unit with W6 as the steam injector and W7, W8, W9, and W10 as the oil producers (Fig. 5.32). At W7 and W8 in the north, yI is classified as Class A geobogy; at W8 and W9 in the south, yI is classified as Class B geobogy. By comparison, the wells penetrating Class A geobody produce a higher oil rate than those penetrating the Class B geobody. This demonstrates that geobodies have critical controls over development performance. Similar conclusions are drawn from other flooding units in the study area, confirming that the parameters selected for geobody classification and evaluation are proper and the results are dependable. Note that a few wells produce extremely low oil rates from yI, even though it is classified as Class A geobody, because yI is waterflooded shortly since production. The wells penetrating the Class B geobody produce at a much lower rate than those penetrating the Class A geobody, but some still produce at a considerable rate. Implementing production measures, such as well pattern improvement and reasonable division of development layers, makes it possible to apply steam flooding to some areas where Class B geobodies are present. However, caution must be exercised in well placement to minimize the loss of heat energy during steam flooding. Also, note that a Class B geobody produces poorer reservoir connectivity between the injector and producer than Class A to avoid the mischoice of production measures.

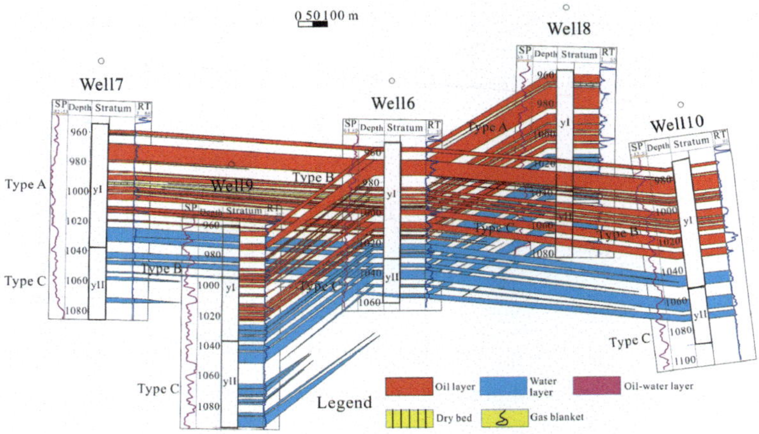

Fig. 5.32 The results of geological body classification and evaluation of Well6 well group in Yulou oil layer in a certain area of western sag of Liaohe Basin. *Injection* Downhole digital representation: oil production/water production/steam injection

5.6.4 Plane Distribution of Geobodies and Its Influence on Development

Figures 5.33 and 5.34 show the plane distribution of geobodies derived from the classification and evaluation. It can be seen the plane distribution of geobodies is strongly governed by sedimentary microfacies. Consistent with sedimentary microfacies, geobogies present a NW–SE distribution, parallel to the provenance. As discussed in the previous section, yI predominately comprises Classes A and B geobodies. Geobody classification matches the developmental effectiveness. Because the geobody controls the steam flooding effectiveness, applying steam flooding recovery to Class C-dominated areas is usually unsatisfactory. The Class A geobody area is a preferred place for steam flooding. By contrast, yII comprises Class C geobodies, despite the Class B and sporadic Class A geobodies in the study area's northeastern part. From the plane distribution maps of geobodies and cumulative liquid production for single wells, a connection exists between high liquid rates, especially high oil rates, and Class B geobodies, demonstrating the reliability of geobody classification. Due to the influences of inherently imperfect and irregular well patterns, water-invaded zones, and technological limitations, some flooding units penetrating Class A geobodies show the performance as Class C in production and must be dynamically adjusted.

Class A geobodies are frequently associated with thick, homogeneous sandbodies with good lateral connectivity that permits uniform advancing of a steam front. Also, the scarcity of barriers/interlayers reduces the loss of heat energy during steam flooding, improving the economic feasibility. At local ultra-high permeability zones where steam channeling is possible, proper measures should be taken to seal the

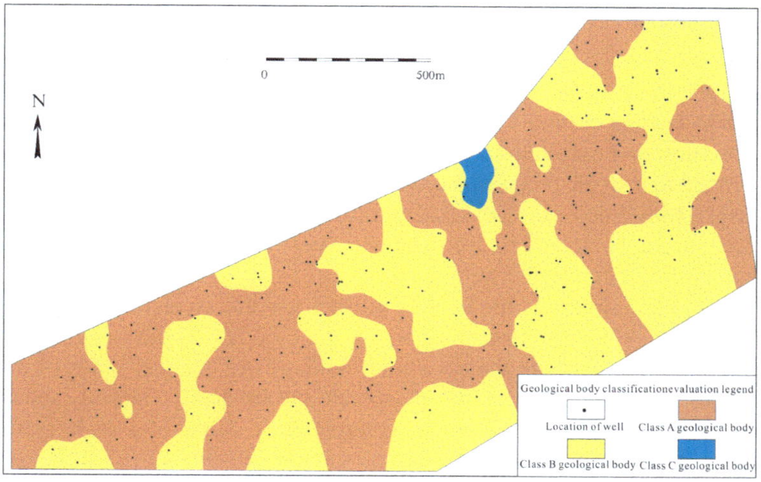

Fig. 5.33 The plane distribution characteristics of different geological body types in yI oil layer group of Yulou oil layer in a certain area of western sag of Liaohe Basin

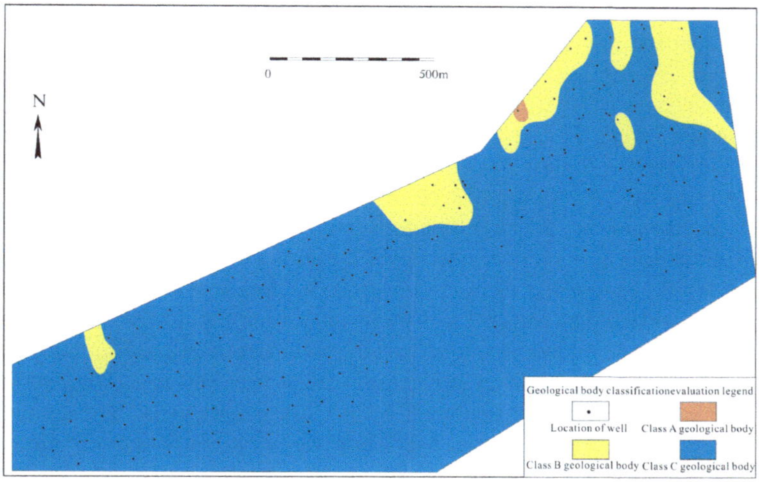

Fig. 5.34 The plane distribution characteristics of different geological body types in yII oil layer group of Yulou oil layer in a certain area of western sag of Liaohe Basin

potential pathways for channeling. Meanwhile, an appropriate foam agent should be used for intralayer and interlayer adjustment and control, or ball-drop selective steam injection is adopted. Furthermore, all wells in a flooding unit should be placed at the same geobody to ensure effective steam flooding. A packer should be set at the boundary between yI and yII to prevent edge and bottom water channeling.

5.7 Summary

Fine stratigraphic division and correlation is the most fundamental task in field development studies. If high-resolution sequence stratigraphy is applied to stratigraphic division and correlation, the guidance of the depositional model is highly recommended to provide precise macroscopic insights regarding the strata development. Also, a stratigraphic division and correlation scheme based on high-resolution sequence stratigraphy should be well-suited to meet the requirements of production practice in mid- to late-stage development.

The sublayer division scheme is determined according to eight principles: (1) at least 50% of the sublayers correspond to one sand body; (2) the sand thickness in a sublayer is not greater than the maximum thickness of channel sands of a single phase, in order to ensure the ability to separate the vertically stacked sand bodies from one another; (3) the well pattern is sufficiently dense at a scale equivalent to or less than a single channel, in order to separate channels of different phases; (4) the boundary of a sublayer usually corresponds to the maximum value on the conductivity log and indicates the location of lake flooding surface; (5) the geological age of a sublayer is approximately 0.03 Ma; (6) the sublayers roughly correspond to the mid- to short-term base-level cycle in high-resolution sequence stratigraphy; (7) the sublayers basically correspond to the fifth-order sequence outlined by Vail in his classic sequence stratigraphy; and (8) the response model of short-term base-level cycles at critical wells can be a reference to delineate the boundary of most sublayers in uncored wells (in other words, it is possible to delineate the short-term base-level cycle in uncored wells).

Integrating well and seismic data forms the basis for high-resolution sequence stratigraphic studies. An accurate large-scale stratigraphic framework is a prerequisite for fine stratigraphic division and correlation. Building the response model of critical wells to varying orders of base-level cycles is a crucial issue to be addressed by high-resolution sequence stratigraphic studies. High-resolution sequence stratigraphic analysis should be coupled with traditional stratigraphic division and correlation methods. High-resolution sequence stratigraphy identifies and correlates varying orders of sedimentary base-level cycles and subdivides bed sets. High-resolution sequence stratigraphy can be combined with traditional methods based on marker bed correlation, sedimentary cycle, log shape variation, and reservoir fluid property variation for cross-validation and supplementation purposes to construct the high-resolution isochronal stratigraphic framework.

Different types of alluvial fans have distinct sediment characteristics, which is significant to reservoir sedimentological analysis and petroleum exploration and development. Alluvial fans can be divided into wet fans and arid fans, depending on factors such as climate, tectonism, and sedimentation.

In the Lower Karamay Formation at the northwestern margin of the Junggar Basin, the mudstones of $S7^4$–$S7^1$ maintain reducing colors, whereas the mudstone of $S6^3$ is oxidized. Extensive scour-filling structures are present in the mid-fan. Conglomerates deposited in this setting are poorly sorted with low textural maturity. $S7^4$–$S6^3$ present as an upward-fining cycle, indicating that the climate changes gradually from humid to arid. The alluvial fan is a transitional type between wet and arid.

The alluvial fan type strongly governs the sandbody's geometry. The wet fan has more intensive fluvial process than the arid fan. The wet fan body is sheet-like, while the arid fan body shows the prominence of debris flow with thick wedge-shaped morphology. During the sedimentation of S7, the alluvial fan in the study area was fluvial-dominated. Sandbodies deposited in different alluvial fan types vary in origin and should be treated differently in field development.

The volcanic reservoirs of the Ying-1 Member in the Xudong area, the Songliao Basin, were formed from multiple volcanic eruptions, comprising various volcanic rocks. The volcanic rocks in the Ying-1 Member can be divided into 10 lithologies, i.e. rhyolite, basalt, rhyolitic breccia lava, rhyolite tuff lava, rhyolitic welded breccia, rhyolitic welded tuff, rhyolitic tuff, rhyolitic volcanic breccia, sedimentary tuff, and sedimentary volcanic breccia. The rhyolite, rhyolitic tuff, and sedimentary volcanic breccia are the most developed. Craters control volcanic rocks, and various volcanic facies are developed at different parts of the volcanic massif. Craters and lithofacies are strongly controlled by faults. The fissure-central eruption pattern is dominant in the study area.

The volcanic reservoirs in the Ying-1 Member are divided into five facies, i.e. volcanic conduit, extrusive, explosive, effusive, and volcanogenic sedimentary, which can be subdivided into 16 subfacies, including volcanic neck, subvolcano, cryptoexplosive breccia, intrazone, meszone, exozone, avalanche, thermal pyroclastic flow, thermal base surge, fallout, top, upper, middle, lower, extraclast-bearing, and retransported. The explosive and effusive facies are most developed.

The physical properties of the Ying-1 Member volcanic rocks are closely related to their lithology and lithofacies. Volcanic facies and subfacies are the primary factors controlling volcanic reservoirs' physical properties, storage space type, characteristics, and variation. The best physical properties can be expected with the avalanche subfacies of the explosive facies with an average effective porosity of 9.00% and an average total permeability of 4.793 mD. Volcanic reservoirs are associated with explosive and effusive facies, except for a few reservoirs related to other facies. Because of the influence and control of volcanic massifs and volcanic eruptions and the subsequent tectonic uplifting and differential compaction, some subfacies are considered favorable, including the avalanche and fallout subfacies of the explosive facies, the exozone and mesozone subfacies of the extrusive facies, the lower, top and upper subfacies of the effusive facies, and the volcanic neck subfacies of the volcanic conduit facies.

Due to the frequent diversion of the underwater distributary channels at fan delta front, the sedimentary permeability barriers vary greatly in the Yulou reservoir in the Liaohe Basin's West Sag. Generally, permeability barriers are relatively developed at underwater distributary interchannel mud and underwater distributary interchannel sand, but rarely at underwater distributary channel and river-mouth bar. In the NW–SE direction, as the distance to the provenance increases, the sediment grains gradually become smaller, and the permeability barrier development is gradually intensified. Affected by the sediment source, the sedimentary permeability barriers with different thicknesses are roughly distributed in form of NW–SE strips. The plane distribution of permeability barriers should be well understood when the thermal recovery mode is converted, so as to ensure the connectivity between injectors and producers for a good flooding effect.

The permeability barriers in the Yulou reservoir are mainly mudstone and silty mudstone, with the effective thickness of 1 m. They complicate the injection–production relationship to reduce the steam flooding efficiency and economic viability. Separate-layer steam injections can be applied to the regions with permeability barriers between sublayers, but may lead to steam channeling in the region without permeability barriers. The permeability barriers yield poor injection-production relationship, impeding the steam flooding efficiency. Thus, the influence of permeability barriers on development should be fully considered when adjusting the injector–producer pattern.

When steam flooding is applied to the Yulou reservoir, the first step is to clarify the effective thickness of oil zone and determine the spatial development of barriers/interlayers by a detailed analysis of sedimentary microfacies and barriers/interlayers on a sublayer basis. Then, the layers to be thermally recovered are divided rationally, and the lower 1/2 or less of the steam injection layer is opened to mitigate the influence of steam overburden. Simultaneously, a steam foaming agent is used for interlayer or intralayer regulation, or the packer is used for separate-layer and ball-drop selective steam injections.

Fine reservoir characterization is a focus of studies associated with field development at the mid- and late-stages. Its core is building a reservoir geological model, in which a primary task is modeling the sedimentary microfacies. Geological modeling is based on sedimentological analysis, a stratification database established using high-resolution sequence stratigraphy, and the geological genetic analysis of structures and reservoirs.

The methods for geological modeling of sedimentary microfacies include the three-dimensional sedimentary microfacies modeling based on horizontal well data and outcrop information, the three-dimensional sedimentary microfacies stochastic modeling based on geologic and seismic data, the stochastic simulation modeling based on marked point process, the modeling based on geological/geophysical log and development dynamic data in combination with geological knowledge, the modeling based on architectural analysis, and the modeling based on cross-well seismic technique. Some key considerations in modeling include the selection of modeling method/technique, selection of modeling software, and principles for facies model validation and selection.

The methods/techniques for modeling sedimentary microfacies have their respective merits and defects. In reservoir characterization, the researchers should consider carefully to choose the strategy and methods/techniques of modeling suitable for the research object, aiming to avoid or even eliminate the defects. In the future, the sedimentary microfacies modeling is believed to develop in the following prospects: (1) Prototype models of various reservoir types and geological knowledge bases are established and improved on the basis of detailed field and subsurface geology studies to provide reliable sedimentary conceptual control and geostatistical guidance to sedimentary microfacies modeling; (2) Attempts are made to explore more optimized algorithms, develop new modeling software based on multivariable geostatistical methods (by applying training image, that is, the reservoir geological mode), and incorporate the researcher's geological knowledge into the modeling process; (3) Various modeling methods/techniques are combined as much as possible to improve the modeling precision and predictability; (4) The latest technologies and research results of multiple disciplines, such as geology, geochemistry, and geophysics, are incorporated into the modeling to improve the modeling precision and efficiency; (5) Stochastic modeling is integrated with deterministic modeling to ensure high precision and predictability; (6) Considering the features of dense well patterns at the mid- to late-stage of development, the development dynamic data are added to the sedimentary microfacies modeling to improve the precision, predictability, and practicability of the model in application to field development.

A geobody functions as the basic unit of steam flooding and it is the material basis for thermal recovery. An in-depth and proper understanding of geobodies directly influences the effectiveness of steam flooding. Several parameters are selected for geobody classification and evaluation, such as effective thickness, net-gross ratio, porosity, permeability, permeability variation coefficient, and interbed frequency. Effective thickness and NGR are the primary parameters, porosity and permeability are the secondary parameters, and permeability variation coefficient and interbed frequency are the reference indexes. Geobodies in the Yulou reservoir in the Liaohe Basin's West Sag are classified into Classes A, B, and C.

The plane distribution of geobodies in the Yulou reservoir is strongly governed by sedimentary microfacies. Classes A and B geobodies are present in yI, whereas Class C geobody dominates yII. Therefore, yI should be the primary target for the steam huff-and-puff to steam injection conversion. Class A geobodies contain thick oil zones (> 35 m), with high NGR (up to 0.45), porosity of over 30%, and permeability mostly > 1500 mD (at least 500 mD). Class C geobodies contain oil zones generally thinner than 5 m, with NGR < 0.1, porosity < 30%, and permeability < 400 mD. Class B geobodies lie between Class A and Class C regarding the properties discussed above and is defined as the secondary target for steam flooding. Class C geobodies are predominately mud zone or water zone, and economically unviable for development.

The Class A geobody area is a preferred place for steam flooding in the Yulou reservoir. At local ultra-high permeability zones where steam channeling is possible, proper measures should be taken to seal the potential pathways for channeling. Meanwhile, an appropriate foam agent should be used for intralayer and interlayer adjustment and control, or ball-drop selective steam injection is adopted. Furthermore, all wells in a flooding unit should be placed at the same geobody to ensure effective steam flooding. A packer should be set at the boundary between yI and yII to prevent edge and bottom water channeling.

References

Bao Z, Ling L, Dongling Z, et al. Depositional system frameworks of the jurassic in Junggar Basin. Acta Sedimentol Sin. 2005;23(2):194–202.

Baosheng R, Fengming L. Reservoir characterization of Dong III Formation, Tangjiahe Developing Area of Beidagang Oilfield, Huanghua Depression. Petrol. Geol. Experiment. 2004;26(1):58–62.

Cao H, Quanshi G, Jinliang T, et al. Discussion of the crosswell seismic data. Progress in Explorat Geophys. 2006;29(5):312–7.

Chen H, Xiaomin Z. Microfacies M odeling in fine reservoir description. Geolog Sci Technol Inf. 2008;27(2):73–9.

Chen S, Hong L, Jianyu Z, et al. The current situation and prospect of crosswell seismic technique. Prog Geophys. 2003;18(3):524–9.

Chen C, Yimei S, Ailin J. Development and application of geological knowledge database for fan-delta front in the dense spacing area. Acta Petrolei Sinica. 2006a;27(2):53–7.

Chen X, Zhu L, Yin Y. Flow unit carved up of sanjianfang formation in III Block in Wenxi area. Petrol Geol Eng. 2007;21(4):36–9.

Chen H, Xiaomin Z, Yanlei D, et al. Sequence stratigraphy analysis and lithologic-stratigraphic reservoir predication in deepwater faulted basins-an example from the Paleogene T Formation of the deepwater C Basin in South China Sea. Oil Gas Geol. 2009;30(5):626–34.

Chen H, Chen C, Shuxian L, et al. Research advances on reservoir pores. Nat Gas Geosci. 2013a;24(2):227–37.

Chen H, Ping J, Danfeng Z, et al. Pore structure of volcanic reservoir classified and distribution characteristics evaluated:Taking M ember 1 of Yingcheng Formation in Xudong Area in Songliao Basin as an example. J Central South Univ (Sci Technol). 2013b;44(4):1453–63.

Chen H, Chunyan L, Jing Z, et al. Research advances of reservoir diageneses. Petrol Geol Oilfield Dev Daqing. 2013c;32(2):1–9.

Chen H, Shuxian L, Zhirui S, et al. Characteristics of conglomerate reservoir architecture of alluvial fan and its controlling effects to reservoir development: Taking alluvial fan reservoir in some area of Northwest Margin of Junggar Basin as an example. J Jilin Univ (Earte Science Edition). 2015a;45(1):13–24.

Chen H, Lei X, Chang L, et al. Classified evaluation of the geological bodies in Yulou Oil Reservoirs of LiaoheBasin. Petrol Geology Oilfield Dev Daqing, 2015b;34(5):45–51.

Chen H, Jiandong M, Jue W et al. Reservoir characteristics of fan delta deposits and its quantitative evaluation: taking Yulou Oil Bearing Sets in a Certain Experimental Area of West Depression in Liaohe Basin as an Example. J Jilin Univ. (Earth science edition), 2017a;47(1):14–24.

Chen H, Chengfang S, Haiyan H, et al. Reflection and countermeasure of fine reservoir description in low oil prices. Geolog Sci Technol Inf. 2017b;36(5):85–91.

Chen Huanqing H, Yongle JJ, et al. Researches on flow unit of the volcanic reservoir in 1st M ember of lower Cretaceous Yingcheng Formation, Xudong area, Songliao Basin. Geology in China. 2011a;38(6):1430–9.

Chen Huanqing H, Yomngle RQ, et al. Lithofacies characteristics of volcanic gas reservoirs and their influence on reservoir physical properties:a ease study of M ember 1 of Cretaceous Yingcheng Formation in Xudong area of the Xushen gas field. Acta Petrologica Et Mineralogica. 2011b;30(1):60–70.

Chi YL, Li B, Su Y, et al. 2D numerical dissection of geological body. Petroleum Geol Oilfield Dev Daqing, 1999;18(1):14–16. (Literature annotation location: 5.6 first stage).

Cross TA. High-resolution, stratigraphic correlation from the perspective of base-level cycles and sediment accommodation. In Anon.Proceedings of Northwestern European sequence stratigraphy congress[S.l.]:[s.n.], 1994; pp. 105–23.

Deng H, Hongliang W, Yongjun Z, et al. High-resolution sequence stratigraphy-principle and application. Beijing: Geological Publishing House; 2002.

Deutsch CV, Journe AG. GSLIB: Geostatistical software library and user guide. London:Oxford University Press; 1996, pp. 167–79.

Dipple GM, Bons P, Oliver NHS. A vector of high-temperature paleo-fluid flow deduced from mass transfer across permeability barriers (quartz veins). Geofluids. 2005;5:67–82.

Duan T, Guangfu W, Peiqing L, et al. Quantitative geological modeling method and application of oil and gas reservoirs. Beijing: Petroleum Industry Press; 2019.

Fan Y, Ji J, Qianyu C, et al. A modified method for storage and reconstruction of three—dimensional geological body model. Geol Explorat. 2019;55(1):203–11.

Feng Y, Yulong W, Xiuchun D, et al. Relationship between intermediate-mafic volcanic facies and their corresponding seismic reflections: a case study from eastern depression of Liaohe Basin. Geophys Prospect Petrol. 2014;53(2):207–15.

Gao J. Lithofacieses and remained oil distribution of pluvial-fan glutenite reservoirs. Petrol Geol Oilfield Dev Daqing. 2011;30(6):62–5.

Galloway WE, Hobday DK. Terrigenous clastic depositional systems-applications to petroleum, coal, and uranium exploration. Spring-Verlag, New York, Inc; 1983.

Gong Q, Geping H, Guohui N, et al. Characteristics of alluvial fan in baikouquan formation of wuerhe oil field in junggar basin and petroleum prospecting significance. Acta Sedimentologica Sinica (2010);28(6):1135–44.

Gong G, Jian G. Subzone correlation of the pluvial fan glutenite reservoirs controlled by datum-level cycle. Petrol Geol Oilfield Dev Daqing. 2011;30(6):18–22.

Guo X, Shoutian C, Xing L. Division and correlation of subatrata in reservoir description: the substrata of dongying formation in Kenxi Oilfield as an example. Geol Sci Technol Inf. 2001;20(2):55–8.

Guo Z, Pujun W, Changhai Y, Yulong H. Relationship between lithofacies and of the logging facies of the volcanic reservoir rocks in Songliao Basin. J Jilin Univ (Earth Sci Edn). 2006;36(2):207–14.

He Z, Pengfei Z, Baiyun et al (2020) 3D solid modeling for complex geological bodies. Geol Expl. 2020;56(1):190–197.

Hen H, Yushuang Z, Qingyin L et al (2006b) Sedimentary microfacies in Chang 6 oil-bearing formation of yanchang formation of Xinghe District in Ansai Oilfield. J Northwest Univ (Nat Sci Edn), 2006b;36(2):295–300.

Hu X, Qihua X, Shenghe W, et al. Advancement of reservoir modeling methods. J Univ Petroleum, China (Edition of Natural Science), 2001; 25(1), 107–112.

Huafeng T, Pujun W, Chuanjin J, et al. Physical model and seismic recognition of concealed volcanic edifices of Yingcheng Formation in Songliao Basin, Cretaceous, NE China. Prog Geophys. 2007;22(2):530–6.

Huang W, Hongmei S, Hailing Z, et al. Characteristics of yingcheng formation'S volcanic reservoir in Xushen gas field in northern Songliao Basin. Acta Petrolei Sinica. 2006;27(supplement):47–51.

Huang Y, Changmin Z, Jun T, et al. Sedimentary microfacies and oil-bearing probability of Lower Kelamayi Formation in Liuzhong District of Kelamayi Oilfield. Nat Gas Geosci. 2007;18(1):67–70.

Hui C. Current status of crosswell seismic technology. Progress in Exploration Geophys. 2002;25(6):6–10.

Ji Y. Petroleum reservoir geology. Beijing: Petroleum Industry Press; 2009. p. 55–134.

Jiang Z. Sedimentology. Beijing: Petroleum Industry Press; 2003a. p. 270–82.

Jiang Z Compiled. Sedimentary.Petroleum Industry Press, Beijing (2003b); 270–282.

Jin Z. Application of svm in identification of high permeability channels. Petrol Geol Oilfield Dev Daqing. 2009;28(6):178–80.

Jing Y, Dameng L, Ruixia W, et al. Identifying volcanic lithofacies of yingcheng form atiom by seism ic attributes in the anda fault•depression of songliao basin. Chinese J Geol. 2009;44(2):595–604.

Joseph DD,Shankar M. Three-dimensional structural model of the painter and east painter reservoir structures,Wyoming fold and thrust belt.AAPG Bullet. 2006;90(8):1171–85.

Kawamoto T, Oil Co. Ltd. T, Sato K. Geological modelling of a heterogeneous volcanic reservoir by the petrological method. SPE59407; 2000: 1–8.

Liu W. Thermal recovery engineering of heavy oil reservoir with steam flooding. Beijing: Petroleum Industry Press; 1997.
Li Q, Meng W, Lichun Z. et al. Conglomerate Oilfield Development. Beijing: Petroleum Industry Press; 1997.
Li Y, Shenghe W, Yanping L, et al. Level characterized of reservoir architecture boundary of delta front channel mouth bar. J Oil Gas Technol. 2007;29(6): 49–52.
Li Y, Zongping S, Qiong L, et al. Prediction of distribution characterss of volcanic facies by using seismic and logging facies. J Mineral Petrol. 2009;29(1):106–13.
Li M, Jiafu Qi, Hengmao T, et al. Cenozoic fault structure and hydrocarbon accumulation in Western Sag, Liaohe Depression. Petroleum Explorat Dev. 2010b;37(3):281–8.
Li G, Huaimin X, Yanqiu L, et al. High-resolution sequence stratigraphy for alluvial fan on lower Karamay formation in 8th zone of northwestern Junggar Basin. J Central South University (Sci Technol), 2010a;41(3):1124–31.
Liu Z, Quanlin X, Junguo D, et al. Formation mechanism and structural model of fault block group oil and gas reservoirs. Beijing: Petroleum Industry Press; 1998b.
Liu J, Shaohua L, Haitao W. Application of sequential indicator random simulation method to study sedimentary microfacies. Petrol Geol Oilfield Dev Daqing. 2003a;22(2):11–46.
Liu L, Wenyan X, Lijuan J, et al. Discussed of volcanic rock cracks formation of Cenozoic in the East Sag in Liaohe Fault-down Basin. Special Oil and Gas Reservoirs. 2003b;10(1):18–21.
Liu G, Lili Z, Liangqian Z, et al. Seismic prospecting for oil and th, @lex eismic prospecting[or O gas on the complex geological bodies. Prog Geophys. 2006a;21(3):683–6.
Liu Y, Zhijun Y, Jinbu L, et al. Application of multi-point geostatistics in sedimentary microfacies modeling of Su 49–01 well area. China Petrol Expl. 2012;4:41–6.
Liu H., Shimin W., Pin . Fracture research the latest progress and research advances in the fracture of the south China sea[J]. Nanhai Yanjiu Yu Kaifa, First Stage: 1998:35–42.
Liu Z, Zhuwen W, Jinghua L, et al. Logging response characteristics and reservoir significance of volcanic rocks in the Eastern Sag of Liaohe Depression. J Jilin Univ (Earth Science Edition), 2018;48(1):285–97.
Liu T, Huaimin X, Jianlin S, et al. Study on the flow units of alluvial fan reservoir in Zhungeer Basin. J. Xi′an Shiyou Univ (Natural Science Edition). 2006b;21(6):24–7.
Longxin M. Stages and characteristic of reservoir description. Acta Petrolei Sinica. 2000;21(5):103–8.
Luo P, Yinan Q, Ailin J, et al. The present challenges of chinese petroleum reservoir geology and research direction. Acta Sedimentol Sin. 2003;21(1):142–7.
Luo J, Tong L, Zhisheng Y, et al. Lithofacies and reservoir quality control factors of volcanics in the Yingcheng Formation in the Shengping Gas Field in Songliao Basin. Oil Gas Geol. 2008;29(6):748–57.
Martín-Izard A, Gumiel P, Arias M et al (2009) Fuertes-Fuente, R. Reguilon. Genesis and evolution of the structurally controlled vein mineralization (Sb-Hg) in the Escarlati deposit (León, Spain): Evidence from fault population analysis methods, fluid-inclusion research and stable isotope data. J Geochem Explorat 100:51–66
Matthew JP, Amanda I, Ellison RD, et al. Analysis and modeling of interm ediate-scale reservoir heterogeneity based on a fluvial point-bar outcrop analog, Williams Fork Formation, Piceance Basin, Colorado. AAPG Bull. 2007;91(7):1025–51.
Mei Z. Sedimentary facies and paleogeography Rebuild. Xi’an: Northwest University Press; 1994.
Miall AD. Architectural-element analysis: a new method of facies analysis applied to fluvial deposits. Earth-Sci Rev. 1985;22:26l–308.
Miall AD. The geology of fluvial deposits, sedimentary facies, basin analysis, and petroleum geology. Berlin: Springer-Verlag; 1996.
Miall AD. Reconstructing the architecture and sequence stratigraphy of the preserved fluvial record as a tool for reservoir development: A reality check. AAPG Bull. 2006;90(7):989–1002.
Miall, AD. Architectural elements sand bounding surfaces in fluvial deposits: anatom y of the Kayenta Formation(Lower Jurassic), Southwest Colorado. Sedimentary Geol 1988;l55:233–62.

Mohammadi S, Hossein Ghazanfari M, Masihi M. A pore-level screening study on miscible/immiscible displacements in heterogeneous models. J Petrol Sci Eng. 2013;110:40–54.

Oriol F, Pau A, Andy G, et al. Best practice stochastic facies modeling from a channel-fill turbidite sandstone analog(the Quarry outcrop, Eocene Ainsa basin, northeast Spain). AAPG Bull. 2006;90(7):1003–29.

Pang W, Mingcai H, Yicai C, et al. Alluvial fan and oil-water distribution characteristics of Xiawuerhe Formation in Well 530field of Karamay Oil Field, Xinjiang, China. J Chengdu Univ Technol (Sci and Technol Edition). 2004;31(5):505–10.

Peng S, Zongsheng L, Jing W, et al. Flow units researches of complexity block reservoir in Shu III area. J Oil Gas Technol. 2009;31(4):170–5.

Ping Z, Chunhui S, Yongbiao Y, et al. The significance and establishment of discriminant function with grain size of stable lacustrine sediment and eolian loess. Acta Sedimentol Sinica, 2008a;26(3):501–7.

Qi L, TR C, Goldstein RH.Geostatistical three-dimensional modeling of oolite shoals,St.Louis Lim estone, southwest Kansas.AAPG Bulletin,2007;91(1: 69–96.

Qin J, Xisheng Z, Chunlai T, et al. Classifying and estimating method of developed blocks in conglomerate oil reservoir. Petrol Geol Oilfield Dev Daqing. 2010;29(3):74–7.

Qinglong D, Yuanqing W, Lihong Z, et al. Formation and distribution of remaining oils in geological bodies of different scales. Pet Explor Dev. 2004;1:95–9.

Qiu Y. Qiu yinan petroleum development geology anthology. Beijing: Petroleum Industry Press; 1997.

Qiu Y, Ailin J. Reservoir geological model for ten years. Acta Petrolei Sinica. 2000;21(4):101–4.

Qiu Y, Ziqi C. Reservoir description. Beijing: Petroleum Industry Press; 1996.

Qingguo Z, Zhidong B, Xinmin S, et al. 2008b. Hierarchical division and origin of single sand bodies in Fuyu oil layer.Fuyu Oilfield. Petrol Expl Devel 2008b; 35(2):157–63.

Ranie L, Elizabeth H. Conceptual model for predicting mud-stone dimensions in sandy braided-river reservoirs. AAPG Bulletin, 2006;90(8):1273–88.

Shu M, Heqing C, Jian W, et al. The flow unit research of braideddelta front subfacies reservoir in E31 oil reservoir of Gasikule Oilfield. Complex Hydrocarbon Res. 2012;5(2):41–4.

Shu P, Rixin D, Yanming Q, et al. Lithologic and lithofacies patterns of volcanic reservoir in Xushen Gas Field. Gas Industry, 2007; 27(8):23–7.

Songjiang Dou Y, Xinghe LC. Flow unit study at middle and late development stages in Northern Dagang oilfield. Oil Gas Geol. 2004;25(1):26–30.

Sosng Z, Wangyan WH, et al. Remaining oil distribution and potential tapping direction in conglomerate reservoir in Karamay oilfield. Petrol Geol Oilfield Dev Daqing. 2008;27(3):44–7.

Sun Y, Xinmin S, Qiquan R, et al. Lithologic and lithofacies characteristics of volcanic rock and their control over reservoirs, Changling Gasfield. Pet Explor Dev. 2009;36(1):68–73.

Shenghe W. Reservoir characterization and modeling. Beijing: Petroleum Industry Press; 2010. p. 157–61.

Shenghe W, Youliang J, Dali Y, et al. Discussion on hierarchical scheme of architectural units in clastic deposits. Geol J China Univ. 2013;19(1):12–22.

Shenghe W, Changfu X, et al. Lithofacies and reservoirs of allluvial fan in the Lower Keramay Formation in the block-6 of Karamay oilfield the Junggar Basin. Oil Gas Geol. 2010b;31(4):463–71.

Steven JI,John WS. A new conceptual model for the structural evolution of a regional salt detachment on the northeast Scotian margin,offshore eastern Canada.AAPG Bullet. 2006;90(9):1407–23.

Tang H, Ji H. Incorporation of spatial characters into volcanic facies and favourable reservoir prediction. SPE90847; 2004:1–11.

Vail PR, Audemard F, Bowman SA, et a1. The stratigraphic signatures of tectonics, eustasy and sedim entology:An over-view[C]// Einsele G,Ricken W,Seilacher A.Cycles and events in stratigraphy. Berlin:Springer-Verlag, (1991); pp. 617–59.

van Leeuwen T, Allen CM, Kadarusman A, et al. Petrologic, isotopic, and radiometric age constraints on the origin and tectonic history of the Malino Metamorphic Complex, NW Sulawesi, Indonesia. J Asian Earth Sci. 2007;29:751–77.

Van Wagoner JC. Siliciclastic sequence stratigraphy in well logs, cores and outcrops. AA PG methods in exploration Series 7. [S.l.]:[s.n.],1990.

Wang X, Lideng G, Weiqi Y, et al. Technical progress of reservoir geophysics. Oil Geophys Prosp. 2006;41(5):606–13.

Wang Y, Yongle H, Qiquan R, et al. Deep layer cracks developed level evaluated of olcanic gas reservoir. Nat Gas Ind. 2007a;27(8):31–4.

Wang Y, Yongle H, Qiquan R, et al. Lithology distinguishing technology research of deep layer volcanic gas reservoir in Xushen Gas Field. J Southwest Petrol Univ. 2007b;29(2):78–81.

Wang Z, Chenghua O, Fengxia L. Study on the lithology discrimination methods for volcanic reservoirs. World Well Logg Technol. 2007c;22(1), 8–11.

Wang P, Zhiqiang F, et al. Compiled. Basin volcanic lithology, lithofacies, reservoir, Gas accumulation and prospecting. Beijing: Press of Science; 2008.

Wei D. Research and application of facies-controlled geological modeling technology for clastic rock reservoirs. Beijing: Science Press; 2016.

Wen W, Xiaotao W. Applied of curvature in complexity geological bodies detected. J Oil and Gas Technol. 2011;33(7):84–7.

Wu S, Zhenkui J, Cangdian H, et al. Reservoir modeling. Beijing: Petroleum Industry Press; 1999.

Wu S, Qihua compiled X. Petroleum reservoir geology. Beijing: Petroleum Industry Press; 1998, 55–104.

Wu H, Fengping Y, Yanguang R et al. Gas reservoir evaluated of Xushen-1 well area of Xujiaweizi Depression in the North of Songliao Basin. Beijing: Petroleun Industry Press; 2002a.

Wu J, Shenghe W, Wei Y, et al. 3D modeling of sedimentary microfacies of minghuazhen formation of neogene in Gangxi development area in Huanghua Depression. J Palaeogeogra. 2002b; 4(4):39–46.

Wu S, Ying L, Zheng F, et al. 3D stochastic modeling of sedimentary microfacies with geological and seismic data. J Palaeogeography. 2003a;5(4):439–49.

Wu S, Junchang W, Shujun L, et al. 3D microfacies modeling of pingqiao horizontal well region of ansai oilfield. Acta Sedimentologica Sinica, 2003b;21(2):266–271.

Wu S, Zhenlin Y, Changfu X, et al. High frequency base-level cycles of alluvial fan and distribution patterns of sandbodies of lower karamay formation (Triassic) in Middle 6th District of Karamayi Oilfield, Xinjiang. Geol J China Univ, 2008;24(2):157–63.

Xia W, Zhanfeng Z, Shiqing C, et al. Geology of oil and gas field development. Beijing: Petroleum Industry Press; 1999.

Xiangyang H, Xiong QihuaShenghe, W, et al. Application of marked point process stochastic simulation to study sedimentary microfacies. J Univ Petrol, China. 2002;26(2):19–30.

Xu Z, Yuming W, Yanming P, et al. Identification and evaluation of Xushen volcan ic gas reservoirs in Daqing. Petrol Exp Dev 2006;33(5), 521–31

Xu Z, Yuming W, Yanming P, et al. Development of volcanic rock reservoirs in the Xushen Gas Field in Daqing. Nat Gas Ind. 2008;28(12):74–7.

Yanshu Yin W, Shenghe ZC, et al. Integrative prediction of microfacies with multiple stochastic modeling methods. Acta Petrolei Sinica. 2006;27(2):68–71.

Yin N, Ji Z, Cungui L, et al. Application of the improved object-based method of high resolution microfacies model to infilled wellblock Su14, Ordos Basin, China. J Chengdu Univ Technol (Sci Technol Edn). 2017;44(1):76–85.

Yin Y, Rui Z, Shenghe W. Integrating data from multiole disciplines in reservoir modeling: a case from the reservoir microfacies of NO.6 District of NO.2 Regin, Gangdong. Nat Gas Geosc. 2007, 18(3):408–11.

Yongwei Shen X, Heng ZC, et al. Flow unit study of upper karamay formation in six middle block in karamay oilfield. J Southwest Petroleum Univ. 2007;29(4):35–55.

Yu J, Futang Y, Huili Q, et al. The geological application of new logging technology in complex lithology formation. J Oil Gas Technol. 2005a;27(2):345–48.
Yu X, Jianyang C, Zhijie Z, et al. Stochastic modeling for characteristics of petroleum reservoir constrained by facies. Earth Sci Front. 2005b;2(3):237–44.
Zhai G, Shihong W, Jiuqiang J. Discussion on block petroliferous geologic body and oil-gas exploration. Acta Petrolei Sinica. 2009;30(4):475–83.
Zhang L. Reservoir connectivity and oil-water relationship of rock dissolved carbonate oil reservoir in tahe oilfield. China Foreign Energy. 2006;11(5):32–6.
Zhang J, Jun X. Development geology. Beijing: Petroleum Industry Press; 2011. p. 60–5.
Zhao W, Caineng Z, Zhiqiang F, et al. Geological features and evaluation techniques of deep-seated volcanic gas reservoirs, Songliao Basin. Pet Explor Dev. 2008;35(2):129–42.
Zhao, H. High resolution sequence stratigraphic correlation and small layer correlation in China. Petroleum Geol Devel Daqing. 2005;24(1):5–12.
Zhao YY, Gu HM, Li ZJ, et al. The Seismic wave field recognition mode of typical geological bodies in the ordovician in Tazhong Area. J Jilin Univ (Earth Science Edition), 2010;40(6):1262–1270, 1286. (Literature annotation location: 5.6 first stage).
Zheng Y, Junchang S, Xiaosong Q, et al. Connotation and evaluation technique of geological integrity of UGSs in oil-gas fields. Nat Gas Ind. 2020;40(5):94–103.
Zheng R, Zhaorong W, Zuowei R, et al. Liaohe sequence stratigraphy and source reservoir caprock assemblage studied of Shahejie Formation of deep stratum in westward in Liaohe Sag in Depression. Multiple Oil-gas Field 1999;4:48–53.
Zheng R, Huaguo W, Fengjie L. High-resolution sequence stratigraphy. Beijing: Geological Publishing House; 2010a.
Zhichao M. Sedimentary facies and paleogeography Rebuild. Xi'an: Northwest University Press; 1994.
Zhou L, Xiong Qihua W, Shenghe. The principles of selecting and testifying facies models in stochastic modelling. Pet Explor Dev. 2001;28(1):68–71.
Zhou L, Limin Z, Guoliang Z, et al. Stochastic facies modeling constrained by high-resolution seismic data. Pet Explor Dev. 2002;29(3):56–8.
Zhu X. Sequence stratigraphy. Dongying: University of Petroleum Press; 2000.
Zhu J. The geological cause analysis of high oil saturation in low porosity and low permeability reservoirs-taking shengli oilfield in Dongying Depression. J Oil and Gas Technol. 2008a;30(3):64–7.
Zhu X. Sedimentary petrology (Fourth Edition). Beijing: Petroleum Industry Press; 2008b.

Uncited References

Chen H. Fracture Study and Its Applied in Oil and gas Field Development. Arab J Geosci. 2019;12:546. (Literature annotation location: Chapter 3).
Chen HL, Tang XY. translated. Secondary reservoir pores in sandstone diagenesis. Beijing: Petroleum Industry Press, 1982. (Literature annotation location: 1.4.3).
Chengjun F, Peng Z, Qi, Bangshen, et al. Recent tectonic stress field at the Shallow Earth's crust near the Tan-lu Fault Zone. Geoscience. 2017;31(1):46–70. (Literature annotation location: 2.2.1).
Chuanhu L, Xuezhong W. Hydrocarbon migration system in complex geological bodies at Chepaizi area, western Junggar Basin. Pet Geol Exp. 2012;34(2):129–33. (Literature annotation location: Part 5.6, first paragraph).
Feng WJ, Wu SH, Xia QY et al. Micro-facies modeling of alluvial fan reservoir based on geological vector information: a case study on the Triassic Lower Karamay Formation, Yizhong Area, Karamay Oilfield, NW China. Geolog J China Universities. 2015;21(3):449–60. (Literature annotation location: 5.5.3.1 Last paragraph)

Fu LY, Xiaofei GF, et al. The dredging and sealing effect of fault on oil and gas. Beijing: Petroleum Industry Press; 2013. (Literature annotation location: 2.2.1).

Fu, LY, Guang ZY, et al. Study on fault sealing capacity. Beijing: Petroleum Industry Press; 2002. (Literature annotation location: 2.2.1).

Fu XD, Wang SN, Zhang L, et al. Sedimentary types and evolution characteristics of Putaohua reservoir in Yushulin-Zhaozhou area, Songliao Basin. Lithologic Reservoir, 2013;25(2):26–30, 35. (Literature annotation location: 3.1 first stage).

Hanqing Z. High resolution sequence stratigraphic correlation and small layer correlation in China. Petroleum Geol Dev Daqing, 2005;2, 24(1):5–12. (Literature annotation location: 5.5.1.2).

Huang W, Shao HM, Zhao HL, et al. Characteristics of yingcheng formation'S volcanic reservoir in Xushen Gas Field in northern Songliao Basin. Acta Petrolei Sinica. 2006;27(supplement):47–51. (Literature annotation location: 5.3 first stage).

Huanqing C., Zhirui, S. Chunyan L, et al. Grain-size analysis of reservoir researches -Taking alluvial fan reservoir in some area of Northwest margin of Zhunger Basin as an example. J Xi'an Shiyou Univeraity (Nat Sci Edition), 2014a;29(6):6–12, 34. (Literature annotation location: 1.4.5. Geostatistics).

Huanqing C, Yingcheng Z, Xingjun G, et al. Types of the alluvial fans of lower Karamay Formation in the Northwestern margin of Zhunger Basin. Petroleum Geology and Oilfield Development in Daqing. 2014b;33(2):6–9. (Literature annotation location: 1.4.5. Geostatistics)

Jingru H, Zhennan Y, Weiming L, et al. Applied geostatistics. Beijing: geological publishing house, 1998. (Literature annotation location: 5.5.3.1(2)).

Lan CL, Wu Jun, Zhang WM, et al. Architectural element analysis for alluvial deposits: principle and applicability. Geol Sci Tech Inf 2002;20(2):37–40. (Literature annotation location: 5.5.2.5 Last paragraph).

Lee K, Gani MR, McMechan GA, et al. Three-dimensional facies architecture and three-dimensional calcite concretion distributions in a tide-influenced delta front. W all Creek Member, Frontier Formation, Wyoming. AAPG Bulletin. 2007;92(2): 191–214. (Literature annotation location: 5.5.2.5).

Li GR, Liu YZ, Li CL. Buried depth inversions of the top and bottom of the magnetic geological bodies. Petrol Geol Oilfield Dev Daqing. 2012;31(6):170–4. (Literature annotation location: 5.6 first stage).

Li CL, Xie CL, Lv QT. Calculating top and bottom depth effect of geological body by applying potential field power spectrum. Petroleum Geol Oilfield Dev Daqing. 1998;17(5):45–8. (Literature annotation location: 5.6 first stage).

Mingli J. Grain-size analysis and its Geological applied. Journal of Oil and Gas Technology. 2009;31(1):161–3. (Literature annotation location: 3.1 first stage).

Pan, Z., Mengjin, C., Zherong, G., et al. Log reservoir and explanation of structureal fractures in volcanic rock reservoir. Acta Petrolei Sinica, 2002;11, 23(6):32–36. (Literature annotation location: 2.3 first stage).

Qing Z, Fubao W, Qiongying C, et al. Grain size distribution characteristics and changes of lop-nur lake sediments during the past 10000 years. J Stratigr. 2009;33(3):283–90. (Literature annotation location: 3.1).

Wu SH, Wang ZL. A new method of non-marine reservoir flow unit study. Acta Sedimentol Sin. 1999;17(2):252–257. (Literature annotation location: 5.4.1 first stage).

Xu, CH, Sun T, S ZQ, et al. Establish the mode chart of alluvial fans and alluvial facies deposition setting by grain size analysis-example as 7th Block Karamay Formation. XinJiang Geology. 2007;25(2):187–91. (Literature annotation location: 3.1 first stage)

Xiaoping Y, Guixia L, Ma, Wenjie, et al. Current situation and developing tendency of sequence stratigraphy. Northwest Geol. 2001;34(2):16–20. (Literature annotation location: 5.5.1.2).

Yang XD, Wang ZX, Guo TZ, et al. Probability cumulative grain-size curves of Triassic flysch deposite in the Bayan Har Mountains area, Qinghai. China Geological Bulletin of China. 2008;27(4):477–90. (Literature annotation location: 3.1 first stage).

Yinan Q, Shuhao X, Fengxiang Y. Continental oil and gas reservoirs in China. Beijing: Petroleum Industry Press; 1997. (Literature annotation location: 5.1.1).

Yuan WF, Chen SY, Zeng CM, et al. Probability cumulative grain size curves in terrigenous of the Tertiary in west Qaidam Basin. J Univ Petrol China. 2005;29(5):12–8. (Literature annotation location: 3.1 first stage).

Zeng LB, He YH, Xiong WL. Origin and geological significance of the cross fractures in the upper triassic yanchang formation, Ordos Basin. Energy Exploration and Exploitation. 2010;28(2):59–70. (Literature annotation location: 1.1 first stage).

Zhaojun L, Pingchang S, Jiangfeng D, et al. Sedimentary Paleocene in characters of fan delta of Tangyuan Fault Depression. J Jilin Univ (Earth Science Edition), 2010;40(1):1–8. (Literature annotation location: 3.1).

Zhengwei L, Li, Wenhou, Long Z, et al. Delationship of grain-size and depositional environment of Chang 10 Formation in Ansai Oilfield. J Shaanxi Univ Sci Technol. 2011;29(1):112–6. (Literature annotation location: 3.1).

Zengzhao F. Sedimentary petrology. Beijing: Petroleum industry press; 1994. (Literature annotation location: 1.4.1).

SPRINGER NATURE

GPSR Compliance

The European Union's (EU) General Product Safety Regulation (GPSR) is a set of rules that requires consumer products to be safe and our obligations to ensure this.

If you have any concerns about our products, you can contact us on ProductSafety@springernature.com

In case Publisher is established outside the EU, the EU authorized representative is:

Springer Nature Customer Service Center GmbH
Europaplatz 3
69115 Heidelberg, Germany

The manufacturer's authorised representative in the EU is Springer Nature Customer Service Centre GmbH, Europaplatz 3, 69115 Heidelberg, Germany. If you have any concerns regarding our products, please contact ProductSafety@springernature.com

Printed and bound by CPI Group (UK) Ltd, Croydon, CR0 4YY
01/10/2026
02225969-0004